An Introduction to Proofs with Set Theory

Synthesis Lectures on Mathematics and Statistics

Editor
Steven G. Krantz, *Washington University, St. Louis*

A Gyrovector Space Approach to Hyperbolic Geometry
Abraham Albert Ungar
2008

.

An Introduction to Proofs with Set   heory

Daniel Ashlock and Colin Lee

ISBN: 978-3-031-01298-3   paperback
ISBN: 978-3-031-02426-9   ebook
ISBN: 978-3-031-00272-4   hardcover

DOI 10.1007/978-3-031-02426-9

A Publication in the Springer series
SYNTHESIS LECTURES ON MATHEMATICS AND STATISTICS

Lecture #35
Series Editor: Steven G. Krantz, *Washington University, St. Louis*
Series ISSN
Print 1938-1743   Electronic 1938-1751

An Introduction to Proofs with Set Theory

Daniel Ashlock
University of Guelph

Colin Lee
Ashlock and McGuinnes Consulting Inc.

SYNTHESIS LECTURES ON MATHEMATICS AND STATISTICS #35

ABSTRACT

This text is intended as an introduction to mathematical proofs for students. It is distilled from the lecture notes for a course focused on set theory subject matter as a means of teaching proofs. Chapter 1 contains an introduction and provides a brief summary of some background material students may be unfamiliar with. Chapters 2 and 3 introduce the basics of logic for students not yet familiar with these topics. Included is material on Boolean logic, propositions and predicates, logical operations, truth tables, tautologies and contradictions, rules of inference and logical arguments. Chapter 4 introduces mathematical proofs, including proof conventions, direct proofs, proof-by-contradiction, and proof-by-contraposition. Chapter 5 introduces the basics of naive set theory, including Venn diagrams and operations on sets. Chapter 6 introduces mathematical induction and recurrence relations. Chapter 7 introduces set-theoretic functions and covers injective, surjective, and bijective functions, as well as permutations. Chapter 8 covers the fundamental properties of the integers including primes, unique factorization, and Euclid's algorithm. Chapter 9 is an introduction to combinatorics; topics included are combinatorial proofs, binomial and multinomial coefficients, the Inclusion-Exclusion principle, and counting the number of surjective functions between finite sets. Chapter 10 introduces relations and covers equivalence relations and partial orders. Chapter 11 covers number bases, number systems, and operations. Chapter 12 covers cardinality, including basic results on countable and uncountable infinities, and introduces cardinal numbers. Chapter 13 expands on partial orders and introduces ordinal numbers. Chapter 14 examines the paradoxes of naive set theory and introduces and discusses axiomatic set theory. This chapter also includes Cantor's Paradox, Russel's Paradox, a discussion of axiomatic theories, an exposition on Zermelo–Fraenkel Set Theory with the Axiom of Choice, and a brief explanation of Gödel's Incompleteness Theorems.

KEYWORDS

set theory, proof, mathematical proofs, logic, functions, relations, integers, cardinal numbers, ordinal numbers, enumerative combinatorics

Contents

Preface

Broadly speaking, mathematics is a subject which concerns itself with problem solving in an abstract manner. If pressed, one might be able to say that mathematics concerns itself primarily with the study of space, quantity, structure, and change.

Students starting university or college may be under the impression that mathematics is primarily concerned about numbers, equations, formula and functions. However, those are merely techniques mathematicians use to solve problem. Mathematics may occasionally (and unfortunately is often) taught in a manner based on mimicry. Students learn to solve several model problems and are then required to recognize which model problem is being encountered and to mimic the standard procedures for the solution of this specific problem. Often this involves plugging the correct numbers in the correctly chosen formula. However, few problems encountered in "real life" are the exact model problems encountered in textbooks. Once the course is finished and the examinations are over this mimicry skill does not have very much lasting value.

That is not to say that there is no benefit from such courses, however a proper under-standing of the reasoning and methods used is necessary to understand when and why the "standard procedures" are valid. If mimicry-based courses have any real value it is only due to the ability of a perceptive and creative student to recognize the more general framework behind the model problems and extrapolate from them.

The primary goal of this course is to encourage and teach students to develop their mathematical perception, creativity, and abstract problem solving ability. We will be doing this by introducing and practicing mathematical proofs. Set theory is a convenient subject matter on which to practice these proofs. It is elementary; it does not require much knowledge of background material, and provides sufficient depth to pose and answer some interesting question through logic and proof. While doing so students will hopefully sharpen their reasoning and become more self-reliant, and critical of mathematical methods that are used.

Daniel Ashlock and Colin Lee
June 2020

Acknowledgments

The authors would like to thank the students that took the course for suggesting many revisions, either directly or through very interesting homework assignments. The authors also gratefully acknowledge the editorial assistance of Wendy Ashlock.

Daniel Ashlock and Colin Lee
June 2020

CHAPTER 1

Introduction and Review of Background Material

Set theory traces its roots to the mathematical investigations of Georg Cantor (1845–1918). Cantor's research uncovered a need to compare the magnitude, or size, of different sorts of infinite sets of numbers. His line of research led to the conclusion that there are all sorts of different types of infinities. Ultimately, thanks to the contributions of a variety of other mathematicians, set theory led to a solid logical foundation for mathematics. It is this later quality which makes set theory a particularly appealing choice of subject matter with which to teach the basic tools of mathematics.

Engineering calculus, possibly the most taken math course in University, is either not a math class or barely a math class depending on how it is taught. It is a course in mathematical techniques that are useful in engineering, physics, some kinds of chemistry, and here and there through the natural sciences. Why is engineering calculus not a math class? This rests on understanding what mathematics is. A professor that taught the first author was fond of saying "Mathematics is the art of avoiding calculation" while calculus is all about performing specific, useful calculations. The goal of mathematics is to expand your intuition and modify the nature of your mind until things that were once opaque and confusing become beautiful and natural—but only some things. Calculations are a side effect of mathematics, not one of its central goals, and this is where calculus classes often fail to be math classes. The term "applied math class" may serve.

Take the **whole numbers**, 0, 1, 2, and so on. Since counting on your fingers is both an approach to the whole numbers and one of the earliest mathematical things that a person does, the whole numbers seem simple. The material in this text will help you see that the whole numbers encompass a complexity that must surpass human imagination. This text and the course it supports take you into the foothills of an endless range of mountains. Endless in a literal and factual, not hyperbolic and poetical fashion.

This text tries to prepare you for a variety of disciples in mathematics, including number theory, abstract algebra, and complex analysis, all of which extend our understanding of the material covered in this text in numerous directions.

While mathematics seeks to expand your mind and intuition to where you can appreciate and use the various disciplines that make up the field, one of the great advances in mathematics was learning to set intuition aside so that the less obvious parts of mathematics could be verified and explored without the traps that form when you rely on intuition. This text introduces this stark discipline within mathematics that is the guardian of correctness: mathematical proofs. The last chapter of the text takes this to an extreme, and discusses axiomatic set theory the foundation of modern mathematics.

As with almost anything, we must begin at the beginning. However, it will be beneficial to be able to communicate in a common language. Before we introduce any new material, we will review a number of mathematical facts and results. This chapter contains a hodgepodge of mathematical information that most undergraduate students should already be familiar with but some of the material may be new. How much of it is new will depend significantly on your educational background. This information will be presented in a relatively informal manner.

1.1   POLYNOMIALS AND SERIES

In mathematics a **polynomial** is a mathematical expression which consists of a variable, coefficients, and a finite number of additions, subtractions, and multiplications. The expression

$$x \cdot x \cdot x + 2 \cdot x \cdot x + 5 \cdot x - 2$$

is a polynomial expression, although it is quite cumbersome to write. For the sake of notational convenience we allow non-negative integer exponents for variables since a finite number of repeated multiplication of variables allows for any positive integer power of a variable. Thus, the expression

$$x^3 + 2x^2 + 5x - 2$$

is also a polynomial. When we group coefficients and powers of a variable together we call it a **term**. So $5x$ is a term, as is $2x^2$.

The **degree** of a polynomial is the highest integer exponent of any term in the polynomial that has a non-zero coefficient. The reason we need to include the non-zero coefficient stipulation is because one can always add additional terms with zero coefficients to any expression without changing the value of the expression. So, the degree of $x^3 + 3x$ is 3 (the exponent of x^3), while the degree of $0 \cdot x^5 + 4 \cdot x^2$ is 2 (the exponent of $4 \cdot x^2$). If there are no variables in a polynomial, the polynomial is said to be of degree zero. This is because $a^0 = 1$ for any non-zero number a.

A **series** is an infinite sum. Series are usually written in the form:

$$\sum_{k=0}^{\infty} a_k$$

The subscript k is used to denote which term in the series we are referring to; the a_k are the terms of the series (these may be coefficients or may include variables). Sometimes, when there are only a finite number of non-zero coefficients we write:

$$\sum_{k=0}^{n} a_k$$

and the series becomes a sum. It's worth noting that while we can actually substitute a value for x into any polynomial, the same cannot be said for every series with polynomial terms. Some series diverge for some or most values of x, for instance $\sum_{k=0}^{\infty} x^k$.

The algebraic properties of the exponential are:

$$a^b \cdot a^c = a^{b+c}$$
$$\left(a^b\right)^c = a^{bc}$$

Examine the following collection of polynomial identities:

$$(1-x)(1+x) = 1-x^2$$
$$(1-x)(1+x+x^2) = 1-x^3$$
$$(1-x)(1+x+x^2+x^3) = 1-x^4$$
$$(1-x)(1+x+x^2+x^3+x^4) = 1-x^5$$
$$(1-x)(1+x+x^2+x^3+x^4+x^5) = 1-x^6$$

These identities work because, when we multiply, all the middle terms cancel, leaving only a 1 and the highest degree term. If we divide by $1-x$ we get the following:

$$1 + x + \cdots + x^n = \frac{1-x^{n+1}}{1-x}$$

Example 1.1

This identity can be used to do certain sums very quickly:

$$1 + 2 + 4 + 8 + 16 + 32 = \frac{1-64}{1-2} = 63$$

$$\sum_{k=0}^{12} 3^k = \frac{1-3^{13}}{1-3} = 797{,}161$$

We now make an application of these identities. Suppose that $|r| < 1$. Then

$$\lim_{n \to \infty} r^n = 0$$

If $|r| < 1$, then we also see

$$\sum_{n=0}^{\infty} r^n = \lim_{n \to \infty} 1 + r + \cdots + r^n$$

$$= \lim_{n \to \infty} \frac{1 - r^{n+1}}{1 - r}$$

$$= \frac{1 - \lim_{n \to \infty} r^{n+1}}{1 - r}$$

$$= \frac{1}{1 - r}$$

It is also easy to compensate for a series that has a constant ratio r between adjacent terms but does not start with 1.

$$a + ar + \cdots + ar^n = a(1 + r + \cdots + r^n) = \frac{a}{1 - r}$$

Example 1.2
This technique works for both finite:

$$2 + 6 + 18 + 54 + 162 + 486 = 2(1 + 3 + 8 + 27 + 81 + 243) =$$

$$2 \cdot \frac{1 - 729}{1 - 3} = 728$$

and infinite:

$$3 + \frac{3}{5} + \frac{3}{25} + \frac{3}{125} + \cdots = \frac{3}{1 - \frac{1}{5}} = \frac{15}{4}$$

versions of these sums.

Definition 1.3 A sum of the form

$$a + ar + \cdots + ar^n$$

or

$$\sum_{n=0}^{\infty} ar^n$$

is called a **geometric series**. The number r is called the **ratio** of the series.

1.2 REPRESENTING NUMBERS

Much of this text will end up revisiting simple things you already have some understanding of and demonstrating the next layer of complexity. You probably already understand place notation for writing decimal numbers. This section reviews this, setting up material in a later chapter on other number bases, but it then goes on to show how different types of numbers have different types of decimal representations. It turns out that the behavior of the digits distinguishes rational and irrational numbers, as we will see in a few pages. There is a subtle interaction between how we choose to represent mathematical objects and what we can understand about them. A good deal of mathematics is the search for an advantageous viewpoint.

Our usual number base for representing numbers is decimal: base 10. When we say 234 we mean

$$2 \times 10^2 + 3 \times 10^1 + 4 \times 10^0$$

Somewhat more oddly, the number 2.34 is

$$2 \times 10^0 + 3 \times 10^{-1} + 4 \times 10^{-2}$$

Both of these are examples of *decimal representations* of numbers.

Definition 1.4   The **decimal representation** of a real number is a, possibly infinite, sequence of digits from the set $\{0, 1, \ldots 9\}$ multiplied by (positive or negative) whole number powers of 10. For reasons explained subsequently, the representation may not end in an infinite sequence of 9s. A number must only use a finite number of positive whole number powers of 10.

The powers of 10 used in the decimal representation of a number form an unbroken sequence of numbers. For 234 this sequence is 2,1,0 while for 2.34 we used 0,-1,-2. We use the digit 0, as a multiplier, to skip positions where a power of 10 does not contribute. Some numbers have fairly complex decimal representations. For instance,

$$\frac{1}{3} = 3 \times 10^{-1} + 3 \times 10^{-2} + 3 \times 10^{-3} + \cdots$$

is a sum with an infinite number of terms. The sequence of numbers that are powers of 10 in this case is "all the negative numbers." There are three classes of decimal representation that we capture with the following three definitions.

Definition 1.5   A number is said to have a **finite decimal representation** or to be a **terminating decimal** if it uses a finite number of powers of 10 in its decimal representation.

Definition 1.6   A number is said to have a **repeating decimal representation** or to be a **repeating decimal** if its decimal representation uses an infinite number of powers of 10 and if the pattern of digits eventually falls into an indefinite repetition of a finite sequence of digits.

Definition 1.7   If a number uses an infinite number of powers of 10 but has no indefinitely repeating pattern of digits, that number is said to be a **non-repeating decimal** or have a **non-repeating decimal representation**.

Example 1.8
We use a bar to denote the repeating part of a decimal representations. The following four numbers are examples of repeating decimals:

(i) $\frac{1}{3} = 0.\overline{3}$ (ii) $\frac{5}{7} = 0.\overline{714285}$

(iii) $\frac{1}{12} = 0.08\overline{3}$ (iv) $\frac{4}{11} = 0.\overline{36}$

Example 1.9
Here are some examples of non-repeating decimals. We will not be able to demonstrate that these are non-repeating decimals until we develop more mathematical power:

(i) $\sqrt{2}$ (ii) $\sqrt[3]{3}$ (iii) π (iv) e

Notice that there is no reasonable method of giving the digits for a non-repeating decimal and so we always use symbols, like those above, to write all of such a decimal. It is often possible to give iterative algorithms of infinite sums for such numbers, e.g., Taylor's theorem in calculus tells us that

$$e = \frac{1}{1} + \frac{1}{1} + \frac{1}{2} + \frac{1}{6} + \frac{1}{24} + \cdots = \sum_{n=0}^{\infty} \frac{1}{n!}$$

or that

$$\pi = \frac{4}{1} - \frac{4}{3} + \frac{4}{5} - \frac{4}{7} + \frac{4}{9} - \cdots = \sum_{n=0}^{\infty} \frac{(-1)^n \cdot 4}{2n + 1},$$

whereas Newton's method tells us that if we let

$$x_0 = 1$$
$$x_{k+1} = x_k - \frac{x_k^2 - 2}{2x_k}$$

that the sequence $n_0, n_1, n_2, \ldots$ converges to $\sqrt{2}$:

$n_0 = 1.0$
$n_1 = 1.5$
$n_2 = 1.4166666667\cdots$
$n_3 = 1.4142156863\cdots$
$n_4 = 1.4142135624\cdots$
$n_5 = 1.4142135624\cdots$

with 11 digits correct by the fourth iteration.

Finding the decimal representation of a fraction can be done by long division:

```
       0.0833...
      -----------
   12)1.000000
        96....
        40...
        36...
        --..
         40.
         36.
         --
        etc.
```

after which you look for a pattern. The fractions $\frac{a}{b}$ where a and $b \neq 0$ are positive or negative whole numbers are called the **rational numbers**. The rational numbers can be characterized by the behavior of their decimal representations. We start with a handy fact.

Proposition 1.10   Suppose that m is a positive whole number with k or fewer digits and let $d = 999\cdots 9$ be a positive whole number whose digits are k 9s. Then $\frac{m}{d}$ is a repeating decimal with the digits of m, possibly including some leading zeros, as the repeating pattern.

In mathematics when we want to demonstrate a claim we use *mathematical proofs*. We will discuss mathematical proofs in great detail starting in Chapter 4, but for now we will simply ask the reader to follow the proof as a demonstration. A mathematical proof is a logical argument which demonstrates a claim (to be more accurate a proof demonstrates that a claim is a logical consequence of certain assumption but we will get into that in Chapters 3 and 4). If the proof of Proposition 1.10 (or later Proposition 1.11) are too difficult to follow at this point, do not worry as we have included them here as reference but they only really need to be understood for Chapter 11. By the time we reach Chapter 11 readers will be a lot more familiar with mathematical proofs.

Proof of Proposition 1.10:

First notice that

$$\frac{m}{d} - \frac{m}{d+1} = \frac{m(d+1) - m(d)}{d(d+1)}$$
$$= \frac{m}{d(d+1)}$$
$$= \frac{1}{d+1}\left(\frac{m}{d}\right)$$

but, since $d + 1 = 10^k$, the fraction $\frac{m}{d+1}$ is a terminating decimal with k digits while dividing $\frac{m}{d}$ by $d + 1$ slides the decimal representation of $\frac{m}{d}$ over by k places. This means that $\frac{m}{d}$ repeats the digits of $\frac{m}{d+1}$, with any necessary leading zeros, indefinitely. This completes the proof. □

The handy fact given in Proposition 1.10 gives us quite a bit of leverage to construct repeating decimals with a given pattern. Consider the example that tells us that $\frac{5}{7} = 0.\overline{714285}$. Proposition 1.10 tells us that

$$\frac{5}{7} = \frac{714{,}285}{999{,}999}$$

This emphasizes that while Proposition 1.10 lets us construct particular repeating decimals, the resulting fractions can be far from being in reduced form.

Proposition 1.11   All rational numbers are terminating or repeating decimals and all terminating or repeating decimals are rationals.

Proof of Proposition 1.11:

Notice that for a terminating decimal like 0.34 we can simply write the digits over an appropriate power of 10:

$$0.34 = \frac{34}{100}$$

and we see all terminating decimals are rationals. Suppose that we have a rational number $\frac{a}{b}$ that does not yield a terminating decimal under long division. Then, when we do long division, the remainders r are in the range $0 \leq r < b$, in particular there are a finite number of distinct remainders. As soon as a remainder repeats the long division falls into a cyclic pattern and so we see that all rationals that are not terminating decimals must be repeating decimals. This argument also shows that the only way that a rational can avoid having a repeating decimal is by having a zero remainder and so being a terminating decimal. It remains only to show that all repeating decimals are rational numbers.

Suppose that r is a repeating decimal. Then $r = q + s$ where q is a terminating decimal and s is a repeating decimal that repeats the same string of digits after some leading zeros. Proposition 1.10 tells us that a repeating string of digits like s is of the form $10^{-d} \cdot \frac{a}{b}$ where a is an integer with the repeating string of digits as its digits, b is a an integer that has only 9s as digits, and d is the number of leading zeros before s starts repeating. We already know q is a rational and we have shown s is so, since the sum of rationals is rational we have that r is rational. This completes the proof. □

We now know that rational numbers are exactly those real numbers that have terminating or repeating decimal representations. We now are in a position to explain our earlier prohibition on decimal representations that end in a endless string of 9s. Ideally there should be only one decimal representation for each real number. The representation we have developed thus far does not have this property *unless* we forbid the terminal 9s. Using Proposition 1.10 we have

$$1 = \frac{9}{9} = 0.99999999\ldots = 0.\overline{9}$$

There are, in fact, an infinite number of these non-unique representations: all repeating decimals with repeating part $\overline{9}$, for example:

$$0.2499999999\ldots = 0.25$$

We solved this problem by declaring that we use only the terminating representation for numbers that could be represented with a repeating decimal with repeating part $\overline{9}$.

1.3 SIMPLE COUNTING

A very old story, one from the Middle Ages, tells the tale of the man that invented the game of chess. A stylized representation of battle with great depth and complexity, chess rapidly became popular with the warrior nobles of the time. The king decided that a reward for great service was merited and asked the inventor what boon he would be granted. The man replied that he would like a grain of wheat on the first square of a chess board, two grains on the second, four on the third, eight grains on the fourth, and so on. The king agreed, because he could not count. At least, not properly. The amount of wheat the king promised was $18,446,744,073,709,551,615$ grains or just under eighteen and one half quintillion grains of wheat. To put this another way, about 1.1 trillion metric tons of wheat.

One of the most basic techniques for counting is **pairing**. We can count a collection of objects by pairing each object in the collection with an object from another collection of objects of a known quantity. For instance, we may count the number of days in the week by going through the days and labeling each with the numbers 1–7. This is the standard method of counting most people are familiar with.

Pairing may be used in other ways. Suppose we want to show that the number of ways of selecting a first, second, and third place from 24 contestants is the same as the number of ways of selecting a class president, vice president, and treasurer from a class of 24 students. If we pair the class president with first place, the vice president with second place, and the treasurer with third place, then we can see that every selection of first, second, and third place from 24 contestants corresponds to a selection of class president, vice president and treasurer from 24 students and vice versa. We do not even need to actually know the value of the quantities to know that both quantities are equal. (In case you were wondering, there are 12,144 ways to make the selections.)

We say that two events are **independent** if the way one happens has no effect on the other. An example of independent events is the number shown on each of two dice that are rolled at the same time but in a manner so that they do not bounce off one another. We say that two events are **mutually exclusive** if the fact that one happened completely prevents the other (in other words the two events cannot both occur at the same time). Having a letter or a digit as the first symbol on a license plate is an example of two mutually exclusive events.

Independent and mutually exclusive events are part of the foundations of probability theory but these two ideas are also useful for counting things once we are aware of the two following facts about the number of possible outcomes of a collection of events.

(i) Independent events multiply.

(ii) Mutually exclusive events add.

Example 1.12
Suppose a pizzeria has 8 meat toppings and 14 veggie toppings available. How many ways are there to top a pizza with one meat and one veggie topping?

Solution:

Since we are not talking about what people might choose or will choose but rather about what they can choose we can count the total number of pizzas by assuming the meat and veggie topping choices are independent. This gives us

$$8 \times 14 = 112$$

possible pizzas.

Example 1.13
Adopt the setup from Example 1.12. How many one-topping pizzas are there?

Solution:

Since a topping is either meat or veggie, there are two mutually exclusive events: meat topping and veggie topping. This then lets us compute that there are

$$8 + 14 = 22$$

one-topping pizzas.

Example 1.14
Adopt the setup from Example 1.12. How many pizzas are there with two veggie toppings? Assume that you can order a double helping of one type of veggie topping.

Solution:

If we assume, as in Example 1.12, that the two choices are independent then there should be $14^2 = 196$ pizzas but there are not. The problem is that a mushroom and onion pizza is the same as an onion and mushroom pizza. When we chose one ingredient first and the other second we created an ordered pair of ingredients but, once the pizza is made, there is no ordering of the toppings.

We cannot just divide the total number of pairs of toppings in half either, because pizzas like the double mushroom pizza appear only once in the list. Pizzas that get counted twice are the ones with two different toppings. There are 14 pizzas that have one topping with a double helping so we can construct a strategy.

1. Compute the total number, $14^2 = 196$ of ordered pairs of toppings.

2. Compute the number of single-veggie, double-helping toppings, 14.

3. Then 196-14=182 is the number of pairs of toppings that are getting counted twice.

4. 182/2=91 is the number of two-topping veggie pizzas with two different toppings.

5. 91+14=105 is the number of two-topping veggie pizzas.

This is a good example of a problem that is not too hard, but it does require you think things through carefully. If you are doing a problem like this and you lack confidence in your technique, then try doing smaller examples—with two meat and three veggie toppings, for example. In this case you can also list the examples by hand to check your method.

1.4   PROBLEMS

Problem 1.15

For each of the following geometric series, state the ratio and compute the sum with the geometric series formula.

(i) $3 + 6 + 12 + 24 + 48 + 96 + 192 + 384$

(ii) $\sum_{k=0}^{12} \frac{2}{3^k}$ (iii) $\sum_{k=0}^{20} 0.9^k$

(iv) $\sum_{k=5}^{14} 2^k$ (v) $\sum_{k=0}^{\infty} \left(\frac{3}{4}\right)^k$

(vi) $\sum_{k=3}^{\infty} \frac{1}{4^k}$ (vii) $\sum_{k=0}^{\infty} \frac{1}{\pi^k}$

(viii) $\sum_{k=0}^{\infty} \left(\frac{x}{x+1}\right)^k$

Problem 1.16

Suppose a ball is dropped from a height of 8 m bounces $\frac{3}{4}th$ as high as it falls. Compute the total vertical distance traveled by the ball.

Problem 1.17

Two bicyclists start 10 km apart. Each rides toward the other at 8 km/hr. A very odd bird starts at one of the cyclists and flies to the other, turns and flies back to the first, and then repeats the round trip until the cyclists pass one another. If the bird is flying 30 km/hr, how far does the bird fly by the time the cyclists pass one another?

Problem 1.18

Using long division, find the decimal representation for each of the following fractions:

(i) $\frac{3}{16}$ (ii) $\frac{7}{125}$ (iii) $\frac{4}{3}$ (iv) $\frac{1}{81}$ (v) $\frac{2}{17}$

Problem 1.19

Find the fractional representation for each of the following decimals. Be sure to put your fraction in lowest terms.

1. $0.\overline{5}$

2. $0.\overline{54}$

3. $1.0.\overline{384615}$

4. $0.132\overline{456}$

5. $0.\overline{0123456789}$

Problem 1.20
Adopt the setup of Example 1.12. Find:

1. the number of pizzas with two meat toppings.

2. the number of pizzas with no more than two toppings.

3. the number of pizzas with three toppings, two of which are meat.

4. the number of pizzas with three veggie toppings.

5. the number of pizzas with no more than three toppings, all of which are different.

Problem 1.21
Suppose a pizzeria offers thick or thin crust made with white, wheat, or gluten-free dough. You can have cheese-in-the-crust or not. They can use pizza sauce, white sauce, spicy marinara sauce, or barbecue sauce for the pizza. They offer 26 toppings and a pizza can have up to 3 toppings. A pizza can also be made half-and-half, but only the toppings can be different on the two halves, not the other choices. How many different types of pizza can be ordered from this pizzeria?

Problem 1.22
Suppose a license plate for a car has either three letters followed by three digits or three digits followed by three letters then how many license plates are possible? Assume no effort is made to avoid spelling rude or unfortunate words or to avoid numbers that are perceived as unlucky or evil.

Problem 1.23
Suppose that we are mixing paint and must use whole cans. If we have three cans of red, two cans of blue, and two cans of yellow paint how many different shades of paint can we mix up? Assume we are mixing in a container with sufficient volume.

Problem 1.24
How many paths from the upper-left corner to the lower-right corner of the grid on the next page if you may only move down or right?

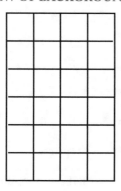

Problem 1.25

Suppose we are filling up an $n \times m$ rectangle with 1×1 squares which can be red or blue. How many different patterns of red and blue can be constructed this way?

CHAPTER 2

Boolean Logic and Truth (Values)

Logic is the basis of proof. The history of logic takes place literally over thousands of years. While valid reasoning has been around even longer, logic, which studies the principles of valid reasoning, has been around since at the very least ancient Babylon. However, some of the logic which is commonly encountered in modern society has only been around for a surprisingly short period of time. Boolean logic, invented by George Boole (1815–1864), the logic on which all of computer science and the modern information age is founded, has only been around from the mid-19th century onward.

Students should not feel "dumb" if they find themselves struggling with some of the finer points of logic which on the surface may seem simple. It has taken the human race thousands of years to discover and understand these "simple ideas."

There are two facets of logic that are important for mathematics. One is used for proofs and argumentation. The other is used to establish the foundations for mathematics. The first is logical argumentation, the principles of deduction cultivated throughout history to codify valid reasoning. This "classical" logic traces its roots to the study of everyday language's use in argumentation. Of concern is the identification and abstraction of those principles used to produce unassailable arguments for the purpose of persuasion. Logic of this kind is concerned with statements of the form:

"If P then Q. P. Therefore Q."

These types of statements are refered to as formal symbolic logic. It is formal in the sense that technically it lacks any reference to interpretation or meaning. The statements symbolized by the letters P and Q could be any sort of proposition. This lack of meaning helps formal symbolic logic achieve great versatility. In mathematics we are primarily concerned with using this sort of logic to achieve **valid arguments**. These are arguments whose form (not content) ensures that the truth of a conclusion is guaranteed to be solely dependant on the truth of our initial assumptions.

A **mathematical proof** is simply the demonstration, by means of a valid logical argument, that the truth of a claim is guaranteed to be the logical consequence of some set of initial premises (assumptions). Thus, in situations where the initial premises are assumed to be true we may be assured of the consequence of the claim.

Every branch of mathematics uses mathematical proofs. These proofs are the means by which mathematicians build up the accepted facts of their subject area. Proofs are sometimes written using formal symbolic logic but they are more often written in language which could be translated into formal symbolic logic.

The second facet of logic important for mathematics concerns itself with the foundations of mathematics. Specifically, we are concerned with formulating a mathematical theory as a system of logic augmented by axioms (these are a specific type of premise). In this aspect of logic we are concerned with essentially "reducing mathematics to pure logic." This facet of logic will be examined in Chapter 14.

2.1   PROPOSITIONS

Definitions are extremely important in mathematics, and they are treated with great care and precision. When we define terms in mathematics, we specify as precisely as possible what is and is not meant by a term. Unlike in everyday language we cannot allow any ambiguity to remain as to whether an object does or does not satisfy a particular definition. Instead of definitions such as defining an elderly person as "a very old person" we require an unambiguous definition such as "a human being over the age of 65."

Definition 2.1   A **proposition** is a declarative statement (or claim) which is either true or false, but not both.

The use of the qualifier "but not both" shows a failure of English as a medium for expressing logical ideas with precision. In logic, if we state that "A is true or B is true" we are happy if A, B or both A and B are true. This **inclusive or** is the standard "or" we adopt in logic and mathematics. In English "or" sometimes has this meaning but it also may mean that only one of A or B is true. In logic we call that an **exclusive or**.

Definition 2.2   The **truth value** of a proposition is the value which the proposition takes, either **True** or **False**.

Example 2.3

1. 2 is an even number.

2. Tokyo is in Japan.

3. $1 + 1 = 7$

4. Apples taste good.

5. Where is the restroom?

6. Burning wood is hotter than ice.

7. If x is an even number then x can be expressed as the sum of two odd numbers.

8. Please pick that up.

The 4th, 5th, and 8th of these "sentences" are not propositions. "Where is the restroom?" is a question. "Apples taste good." is a declarative statement but it is subjective as to whether it is true or false. By itself it is not a proposition unless it is further qualified, such as "Andrew thinks apples taste good." "Please pick that up." is a command.

The remainder of these "sentences" are propositions. Note that propositions may be used to make claims about abstract mathematical facts (such as "2 is an even number."), implications (something of the form "if A then B"), or facts which require a great deal of context to evaluate (such as "Burning wood is hotter than ice").

The 1st, 2nd, 6th, and 7th of the "sentences" are "true" in the sense that they correspond to reality. However this notion of truth does not necessarily have anything to do with their truth values in a logical sense. This will be discussed in greater depth later, but for now it's important to keep in mind that at times "corresponds with reality" may not be the most useful way to think about "truth" in mathematics and logic.

In mathematics we often encounter declarative sentences which are formed by modifying simpler propositions with **connectives**, words such as *and, or, not, if . . . then* or *if and only if*.

Definition 2.4   A **compound proposition**, also called a **composite proposition** is a proposition composed of various simpler propositions by means of connectives.

Definition 2.5   A **simple proposition**, also called a **primitive proposition** is a proposition which is not a compound proposition.

Example 2.6
Consider the following propositions:

1. The sky is blue, and the birds are singing.

2. We did not both go to the movie and go to the theater.

3. Mary, John, and Dennis went to lunch together.

4. An integer is a sum of two odd numbers or it is an odd number.

5. If a function is differentiable then that function is continuous.

In the first example, "the sky is blue" is a simple proposition and "the birds are singing" is a simple proposition, they are made into a compound proposition by means of the connective "and". In the second example, "we went to the movie" and "we went to the theater" are simple propositions. They are joined by the "and" connective, but then everything is **negated** by means of the "not" connective. The entire compound proposition is of the form **NOT (A AND B)**. The third example is a simple proposition. Although it contains an "and" it is expressing one piece of true or false information—that Mary, John, and Dennis *all* went to lunch *together*. The fourth and fifth examples are compound propositions.

2.2   LOGICAL OPERATIONS AND TRUTH TABLES

The values **true** and **false** form the space of logical values, also called *Boolean values* after George Boole who developed the area as a type of mathematics. There are a number of operations on Boolean values. Since there are only two possible values in the Boolean system, operators can be specified by listing all the value. Formally, we represent connectives by means of *Boolean operators*.

Definition 2.7   Any two propositions P and Q can be combined by an "and" to form the **conjunction** of P and Q. Written $P \wedge Q$, or (P AND Q) and read as "P and Q." The proposition $P \wedge Q$ is true when both P is true and Q is true, otherwise it is false.

Definition 2.8   Any two propositions P and Q can be combined by an "or" to form the **disjunction** of P and Q. Written $P \vee Q$, or (P OR Q) and read as "P or Q." The proposition $P \vee Q$ is false when P is false and Q is false, otherwise it is true.

Definition 2.9   The **negation** of a proposition P is written $\neg P$ or (NOT P). It is read as "NOT." NOT P is true when P is false and false when P is true.

These three logical operations can be combined to produce very complex compound propositions. In fact, it is possible to produce *any* compound proposition made up of a finite number of simple propositions using only these operations. However, before we deal with increasingly complicated compound propositions we'll introduce a convenient method for specifying and analyzing them. It is perhaps worth mentioning that we've used P and Q to symbolically refer to propositions, we are using these as **logical variables** or **Boolean variables**—variables which only take on truth values. We specify each simple proposition with a separate variable. The truth value of a compound proposition depends entirely on the truth values of the variables it is composed of.

Just like with mathematical operators there is a convention about the order in which to evaluate logical operators when parentheses are not used. The negation operator takes precedence followed by the conjunction and then disjunction operators. In other words, $\neg A \wedge B \vee C$ is actually the same as $((\neg A) \wedge B) \vee C$.

Definition 2.10   A **truth table** is a specification of an operation by listing all the possible (truth) values that could be presented to the operation and the result of applying the operation to them.

The three standard Boolean operations are given below. The first two, AND and OR, take two truth values and return one. The last, NOT, operates on a single truth value. The operations are applied to the truth values of propositions A and B (or just A if only one proposition is needed). The table lists all possible values for truth values of A and B and the result of applying the operation. We use T for true and F for false.

AND		
A	B	result
F	F	F
F	T	F
T	F	F
T	T	T

OR		
A	B	result
F	F	F
F	T	T
T	F	T
T	T	T

NOT	
A	result
F	T
T	F

Truth tables are a method for specifying the (possibly complicated) truth value assignments of composite propositions. We will return to them but first need to discuss a special compound proposition.

2.3   IMPLICATION

Mathematics involves a lot of statements of the form "IF P THEN Q," "P is a sufficient condition for Q," "Q is a necessary condition for P" or "P IMPLIES Q." All of these statements are equivalent, they are all an **implication**. The P is the **antecedent** of the implication and the Q is the **consequent** of the implication. Implications are sometimes called **conditionals**. There is a formal definition we use for implication in mathematics.

Definition 2.11   The **conditional** P implies Q, written symbolically as $P \rightarrow Q$, which is read as "P implies Q," is the compound proposition which is false when P is true and Q is false but is true otherwise.

The truth table for implication, denoted as IMPLIES or via the symbol $\rightarrow$, is given below.

IMPLIES		
A	B	result
F	F	T
F	T	T
T	F	F
T	T	T

Students often have a difficult time with this definition of implication. Somehow the statements "If pigs can fly then the sky is blue." and "If pigs can fly then there are exactly two people on earth." are both true and this seems to be a cause for concern. Usually this arises due to a difficulty accepting what the truth value of an implication *should* be when the antecedent is false. It seems to run counter to some of their intuition. Rather than simply referring them to the fact that we actually defined it to be false only when the antecedent is true and the consequent is false, it is perhaps more useful to think of it in terms of a contract.

Example 2.12
Alex makes a contract with Sasha. *If it rains tomorrow then Alex will buy Sasha an umbrella.* Now, let's examine when exactly Alex would be breaching his contract (i.e., breaking a promise). Suppose it rains tomorrow and Alex does in fact buy Sasha an umbrella. Clearly, Alex would be fulfilling his part of the contract. Suppose it does not rain tomorrow. Then Alex is not required to buy Sasha an umbrella. Whether he does or does not, Alex would still be fulfilling his part of the contract. However, if we suppose it rains tomorrow and Alex does not buy Sasha an umbrella then this would be a circumstance in which Alex is breaching his contract (i.e., breaking a promise).

We hope that the contractual interpretation helps explain the issue. In any event in logic we deal with what is known as **material implication**, the conditional just mentioned which is true in any case except when the antecedent is true while the consequent is false. We do not deal with any *causal relationships* or *casual implication* the "standard" use of "if ... then" which students seem to get confused with material implication. In a causal implication of the form "if P then Q" the P is the cause and the Q is the consequence, somehow P being true causes Q to be true. For example, "If you wake up early then you will be able to see the sunrise." and "If you park in the driveway then I won't be able to get my car out of the garage." are causal implications. It so happens that material implication and causal implication share the same everyday language of "if-then" but that is all. In logic we stick with material implication.

There is another common type of statement that occurs frequently in mathematics; they are of the form P if and only if Q. These are called **biconditionals**.

Definition 2.13   The **biconditional** P if and only if Q, written symbolically as $P \leftrightarrow Q$, which is often also written as P iff Q, is the compound proposition which is true when P and Q both have the same truth value and false when their truth values differ. P iff Q is true when P and Q are both true or both false and false otherwise.

The proposition $P \leftrightarrow Q$ is the same as saying "P is a necessary and sufficient condition for Q." The biconditional statement is used to capture the notion of *equivalence* or more accurately **logical equivalence**. When we assert that a statement $P \leftrightarrow Q$ is true we are stating that P and Q have the exact same truth value, even when P and Q are compound propositions themselves. However, since the compound proposition $P \leftrightarrow Q$ may itself be true or false, we use the convention that $P \equiv Q$ is used to denote *logical equivalence when the assertion that P and Q are logically equivalent is true*. In other words, $P \equiv Q$ exactly when P and Q are logically equivalent and thus have the same truth tables.

We note that $P \rightarrow Q \equiv \neg P \vee Q$ and that $P \leftrightarrow Q \equiv (P \rightarrow Q) \wedge (Q \rightarrow P)$.

$P \rightarrow Q$		
P	Q	result
F	F	T
F	T	T
T	F	F
T	T	T

$\neg P \vee Q$		
P	Q	result
F	F	T
F	T	T
T	F	F
T	T	T

$P \leftrightarrow Q$		
P	Q	result
F	F	T
F	T	F
T	F	F
T	T	T

$(P \rightarrow Q) \wedge (Q \rightarrow P)$		
P	Q	result
F	F	T
F	T	F
T	F	F
T	T	T

Definition 2.14   The **contrapositive** of the implication $P \rightarrow Q$ is the implication: $\neg Q \rightarrow \neg P$.

Definition 2.15   The **converse** of the implication $P \rightarrow Q$ is the implication: $Q \rightarrow P$.

Note that the contrapositive of an implication has the exact same truth table as the original implication, so these two implications are *logically equivalent*, whenever one is true the other is true and whenever one is false the other is false.

The converse of an implication *does not* have the same truth table as the original implication. They are not logically equivalent. A common mistake people make when using logic is to confuse the two.

$P \to Q$				$\neg Q \to \neg P$				$Q \to P$		
P	Q	result		P	Q	result		P	Q	result
F	F	T		F	F	T		F	F	T
F	T	T		F	T	T		F	T	F
T	F	F		T	F	F		T	F	T
T	T	T		T	T	T		T	T	T

2.4 TAUTOLOGIES AND CONTRADICTIONS

Some compound propositions are always true or always false regardless of the truth value of their logical variables. These are **tautologies** and **contradictions**.

Definition 2.16   A **tautology** is a compound proposition which has a truth value of true for all possible combinations of truth values for its logical variables.

Definition 2.17   A **contradiction** is a compound proposition which has a truth value of false for all possible combinations of truth values for its logical variables.

Example 2.18
$A \wedge \neg A$ is a contradiction.
$A \vee \neg A$ is a tautology.
$(A \wedge B) \to A$ is a tautology.

In a truth table the results column of a tautology is all "T" and the results column of a contradiction is all "F".

2.5 LAWS OF THE ALGEBRA OF PROPOSITIONS

Propositions and the three main connectives ($\wedge$, $\vee$, and $\neg$) satisfy a variety of algebraic laws which are listed subsequently. Here, T and F may refer to truth values (true and false), when evaluating the formula or they may equivalently refer to tautologies and contradictions (T and F, respectively).

Idempotent Laws	
$P \lor P \equiv P$	$P \land P \equiv P$
Associative Laws	
$(P \lor Q) \lor R \equiv P \lor (Q \lor R)$	$(P \land Q) \land R \equiv P \land (Q \land R)$
Law of Double Negation	
$\neg\neg P \equiv P$	
Commutative Laws	
$P \lor Q \equiv Q \lor P$	$P \land Q \equiv Q \land P$
Distributive Laws	
$P \lor (Q \land R) \equiv (P \lor Q) \land (P \lor R)$	$P \land (Q \lor R) \equiv (P \land Q) \lor (P \land R)$
Identity Laws	
$P \land T \equiv P$	$P \land F \equiv F$
$P \lor T \equiv T$	$P \lor F \equiv P$
Complement Laws	
$P \lor \neg P \equiv T$	$P \land \neg P \equiv F$
$\neg T \equiv F$	$\neg F \equiv T$
De Morgan's Laws	
$\neg(P \land Q) \equiv \neg P \lor \neg Q$	$\neg(P \lor Q) \equiv \neg P \land \neg Q$

While the majority of these laws are probably self-evident (like the Idempotent Laws and the Law of Double Negation), some such as the Distributive Laws and De Morgan's Laws may not be. However, all of these laws can be verified via truth tables.

2.6   ADDITIONAL EXAMPLES

Example 2.19

Determine the simple propositions in the following, possibly compound, propositions:

1. The rocks at the shoreline are smooth but the ones further into the forest are rougher.

2. If Susan and Nala are able to catch the 6:15 train, they need to leave campus by 5:55.

3. Neither Chris nor Matthew took a summer class this year.

Solution:

1. "The rocks at the shoreline are smooth," and "the rocks further into the forest are rougher" are the simple propositions.

2. "Susan and Nala are able to catch the 6:15 train" and "Susan and Nala need to leave campus by 5:55" are the simple propositions.

3. Either "Chris took a summer class this year" and "Matthew took a summer class this year" are the simple propositions (which are then negated) or "Chris did not take summer class this year" and "Matthew did not take a summer class this year" are the simple propositions.

Example 2.20
Using the laws of the algebra of propositions, show that
$\neg(A \vee B) \vee \neg A \equiv \neg A$.

Solution:

$\neg(A \vee B) \vee \neg A$
$\equiv (\neg A \wedge \neg B) \vee \neg A$ by De Morgan's Law
$\equiv (\neg A \wedge \neg B) \vee (\neg A \wedge T)$ by the Identity Law
$\equiv \neg A \wedge (\neg B \vee T)$ by the Distributive Law
$\equiv \neg A \wedge T$ by the Identity Law
$\equiv \neg A$ by the Identity Law

2.7 PROBLEMS

Problem 2.21
Determine which of the following sentences are propositions:

1. Behold the dawn's early light.

2. Four times twelve is forty four.

3. Frank has a chipped tooth.

4. Tomorrow's sunset will be magnificent.

5. After class ends today, I have a forty-five minute walk home.

6. If Joe goes to the movies, and Anya goes to the coffeeshop, then I'll have the apartment to myself.

Problem 2.22
Determine the simple propositions in the following, possibly compound, propositions:

1. If Gia or Mary scores higher than Todd on this test, then Todd will no longer have the highest marks in the class.

2. The equation $x^2 + 1 = 0$ has no rational solution.

3. If Bill went upstairs, then he needed to take the stairs or take the elevator.

4. I cannot go to both the new horror movie and the football game on Tuesday.

5. Since Melissa moved to Japan and Christopher moved to Italy, I'm the only one left in Canada.

Problem 2.23
Let P, Q, and R be the following propositions:
P: There are mountain lions in the area.
Q: Hiking is safe on the trail.
R: There is a forest fire in the area.
Write the following propositions using P, Q, R, and logical connectives.

- Hiking is safe on the trail, so there isn't a forest fire or mountain lions in the area.

- If there are mountain lions in the area, then there isn't a forest fire in the area.

- There isn't a forest fire in the area, but hiking isn't safe on the trail.

- If there is a forest fire in the area, or there are mountain lions in the area, then hiking is not safe on the trail.

- In order for hiking to be safe on the trail, it is necessary that there isn't a forest fire in the area.

- Hiking is safe on the trail, if there aren't mountain lions in the area.

Problem 2.24
The exclusive-or on pairs of Boolean values is true if exactly one of the truth values on which it is operating is true. Give a truth table for exclusive-or, which is denoted XOR.

Problem 2.25
Determine which of the following conditionals is true and which is false.

1. If an integer can be written as the sum of two odd numbers, then it is an even number.

2. If $1 + 1 = 4$, then the y-intercept of $y = mx + b$ is b.

3. Pigs can fly, if ice is colder than steam.

4. If $1 + 1 = 2$, then $3 + 3 = 3$.

5. Mary has a cat and Steve doesn't have one if and only if it is not the case that if Mary has a cat then Steve has a cat.

Problem 2.26
State the contrapositive and the converse of the following implications (and specify which is which).

1. If it is hot outside, then the dog hides under the deck.

2. $A \vee B \to C$

3. Whenever an integer can be expressed as the sum of two odd numbers it is divisible by two.

4. If the earth rotates around the sun, then the sun rotates around the earth.

Problem 2.27
Using a truth table, demonstrate that $(P \wedge Q) \to P \vee Q$ is a tautology.

Problem 2.28
Using a truth table, demonstrate that $R \to (P \wedge Q) \equiv (R \to P) \wedge (R \to Q)$.

Problem 2.29
Using the laws of the algebra of propositions, show that
$(\neg A \wedge B) \vee (\neg A \wedge \neg B) \equiv \neg A$.

Problem 2.30
Using the laws of the algebra of propositions, show that
$\neg(A \vee C) \wedge B \equiv \neg(A \vee \neg B \vee C)$.

CHAPTER 3

Quantified Predicates, Rules of Inference, and Arguments

This chapter introduces quantified predicates and rules of inference. Combined with the previous chapter, a firm grasp of these concepts are the major tools needed for most sorts of logical arguments.

3.1   PREDICATES AND QUANTIFIERS

So far we've discussed what is known as **propositional logic**. The objects under consideration were propositions, declarative statements that are true or false. By and large, most common uses of symbolic logic may be intuitively grasped in terms of a firm understanding of propositional logic. However, in certain circumstances it's necessary to put variables in our propositions.

We would like to logically argue about statements such as "If $x(x + 1) > 0$, then $x > 0$," for example.

Statements involving variables make it difficult to determine the truth value of propositions, as the value of the variable may affect whether the statement is true or false. Consider the following statement: "x is less than 9." Whether the statement is true or false clearly depends on the choice of x. The statement has two parts; the first of which is the variable x. The part of the statement "is less than 9" is called the **predicate**, and refers to the property that the subject of the statement may (or may not) have. We can denote the statement "x is less than 9" by the **propositional function** $P(x)$. Here, P refers to the predicate and x is the variable. Note that once an actual value has been assigned to x then $P(x)$ becomes a proposition with a definite truth value.

Definition 3.1   A **propositional function** $P(x)$ is a declarative statement with a variable x which becomes an actual proposition whenever a (valid) value is substituted in place of x. In this situation x is the variable and P is the **predicate**.

Whenever a propositional function's variable is assigned a value the function becomes a proposition. However, there is another important way in which we want to be able to create propositions from propositional functions. We want to be able to discuss $P(x)$ when x is considered as a variable. For this we need **quantifiers**.

Many mathematical statements make some sort of claim about a property that is true for all values of a variable in some particular domain. That domain is called the **domain** or **universe of discourse**. For example, in "Every even number x can be written as a sum of two odd numbers." the universe of discourse is the even numbers.

Definition 3.2   The **universal quantification** of a propositional function $P(x)$ is the proposition: "For all values of x in the universe of discourse, $P(x)$ is true."

The notation $\forall x,\ P(x)$ is used for the universal quantification of $P(x)$. The symbol $\forall$ is called the **universal quantifier**, and is read as "for all" or "for every."

Example 3.3

Let $P(x)$ be the propositional function "$x + 1 > 0$." If the domain of discourse consists of all positive numbers, then $\forall x,\ P(x)$ is true, since every x is itself greater than 0. If the domain of discourse consists of all real numbers, then $\forall x,\ P(x)$ is false, since when $x = -2$ then $x + 1 < 0$. Finally, if the domain of discourse is all the cows in farmer Brown's field, then $P(x)$ is nonsensical (and is not in fact a propositional function *in that particular domain of discourse*).

The final portion of the previous example should clarify what was meant by the qualifier of a "(valid) value" in the definition of a propositional function. We simply mean a value from the domain of discourse.

The proposition $\forall x,\ P(x)$ is true when $P(x)$ is true for every x in the universe of discourse. This means that $\forall x,\ P(x)$ must be false when there is in fact an x in the universe of discourse for which $P(x)$ is false. It thus becomes clear that we will need to be able to discuss the existence of *counterexamples* to statements of the form $\forall x,\ P(x)$.

Definition 3.4   The **existential quantification** of a propositional function $P(x)$ is the proposition: "There exists a value of x in the universe of discourse, such that $P(x)$ is true."

The notation $\exists x,\ P(x)$ is used for the existential quantification of $P(x)$. The symbol $\exists$ is called the **existential quantifier**, and $\exists x\ P(x)$ is read as "there exists x such that $P(x)$."

Example 3.5

Let $P(x)$ be the propositional function "x is a multiple of 3." If the domain of discourse is the integers, then $\exists x,\ P(x)$ is true since $x = 3$ would be an example that demonstrates the claim.

If the domain of discourse consists only of the numbers 1, 2, and 4, then $\exists x,\, P(x)$ is false since $P(x)$ is false for $x = 1, 2$, and 4.

It should be clear from the definitions that if $P(x)$ is a propositional function then:

$$\neg(\forall x,\, P(x)) \equiv \exists x,\, (\neg P(x))$$

and

$$\neg(\exists x,\, P(x)) \equiv \forall x,\, (\neg P(x))$$

With the addition of universally and existentially quantified propositional functions to the simple and compound propositions from the previous chapter we are able to discuss and symbolically specify logical statements. Now that it's possible to represent claims as precise logical statements the question becomes, how can we use these ideas to make logical arguments?

3.2   ARGUMENTS

What is a logical argument? In everday language when someone advances a rational or "logical" argument they are generally supporting claims they have made with (hopefully) solid factual evidence. Perhaps they are able to amass a large volume of evidence which taken together makes it seem highly probable that the claim they are making is true. (True in the sense that it conforms to reality.) This sort of "rational argument" may be useful for the purposes of persuading other people that a particular claim is true. However, such a person is using **inductive reasoning**; their premises (the evidence) is used to supply *probable* conclusions. There is no guarantee that the conclusion is true, even when all the premises (the factual evidence) is true.

In mathematics, inductive reasoning isn't good enough, we need to guarantee that the conclusion must be true when the premises are true. What we care about is **deductive reasoning**.

We cannot guarantee the truth of any particular simple proposition, even universally agreed upon facts, such as "the sky is blue" or "it does not literally rain cats and dogs." However, if we can take it as a given that the basic rules of logic work as intended, such as that conjunction and disjunction work in the way we need "and" and "or" to work, then we can produce composite propositions which are tautologies. Recall that tautologies are true, in the truth value sense, purely by virtue of their form.

Example 3.6
Consider the proposition $P \wedge Q \to P \vee Q$ which is a tautology. Implications of the form $A \to B$ are false when A is true and B is false. So since $P \wedge Q \to P \vee Q$ is a tautology (and always true), whenever the antecedent $P \wedge Q$ is true it is guaranteed that the consequent $P \vee Q$ cannot be false (and hence must be true).

Definition 3.7

An **argument form** is an implication of the form $(P_1 \wedge P_2 \wedge \ldots \wedge P_n) \rightarrow Q$, where the P_i are called the **premises**, and the Q is called the **conclusion**.

In the context of argument forms the final implication is usually read as "therefore," thus $A \wedge (A \rightarrow B) \rightarrow B$ is read as "A and A implies B, therefore B."

Definition 3.8   An argument form $(P_1 \wedge P_2 \wedge \ldots \wedge P_n) \rightarrow Q$ is called **valid** (or a **valid argument**) when $(P_1 \wedge P_2 \wedge \ldots \wedge P_n) \rightarrow Q$ is a tautology.

In other words, an argument form is a valid argument whenever it's the case that if the premises are all true then the conclusion must be true. Valid arguments are examples of correct deductive reasoning. A valid argument cannot guarantee the truth of its premises, however a valid argument can demonstrate that if the premises are all true then the conclusion is *guaranteed* to be true.

Valid arguments are essential for mathematical proofs. In a certain sense mathematical proofs are just a series of valid arguments which taken together constitute a more complex valid argument. We will discuss this in greater depth in the next chapter.

It is important to note that any propositions may be substituted in place of the simple propositions which appear in a valid argument and the resulting argument is still valid. This is because tautologies are true at all times, due to their form not the particular content of the simple propositions which make up a valid argument.

Example 3.9

Consider the argument form $P \rightarrow P \vee Q$, which is a tautology. It corresponds to the argument "It is a Saturday, therefore it is a Saturday or it is the first of the month." as well as the argument "Alice chased the rabbit, therefore Alice chased the rabbit or Alice fell asleep by the tree." Also, one may substitute the compound proposition $A \wedge B$ in place of P and the resulting proposition $(A \wedge B) \rightarrow (A \wedge B) \vee Q$ is still a valid proposition.

3.3   RULES OF INFERENCE

There are several valid argument forms that occur with such frequency when people attempt to deduce conclusions that they have their own names. They are called the **rules of inference**. In fact, although we present them as *valid argument forms* they were known long before Boolean logic was ever invented. They form the basis for deductive logic, the art of demonstrating that a conclusion follows logically from a set of initial assumptions (the **premises**). See Table 3.1 for the full list of rules of inference.

The tautology $((P \rightarrow Q) \wedge P) \rightarrow Q$ is the first rule called **Modus Ponens**. Modus Ponens and the other rules of inference are generally written via the following convention:

$P \rightarrow Q$
P
$\therefore Q$

The symbol $\therefore$ is read as "therefore," it is used to indicate the implication portion of the argument form.

Table 3.1: Rules of Inference

Rule	Tautology	Name
$P \rightarrow Q$ P $\therefore Q$	$((P \rightarrow Q) \wedge P) \rightarrow Q$	Modus Ponens
$P \rightarrow Q$ $\neg Q$ $\therefore \neg P$	$((P \rightarrow Q) \wedge \neg Q) \rightarrow \neg P$	Modus Tollens
P $\therefore P \vee Q$	$P \rightarrow (P \vee Q)$	Addition
$P \wedge Q$ $\therefore P$	$(P \wedge Q) \rightarrow P$	Simplification
P Q $\therefore P \wedge Q$	$((P) \wedge (Q)) \rightarrow P \wedge Q$	Conjunction
$P \vee Q$ $\neg P$ $\therefore Q$	$((P \vee Q) \wedge \neg P) \rightarrow Q$	Disjunctive Syllogism
$P \rightarrow Q$ $Q \rightarrow R$ $\therefore P \rightarrow R$	$((P \rightarrow Q) \wedge (Q \rightarrow R)) \rightarrow (P \rightarrow R)$	Hypothetical Syllogism
$P \rightarrow Q$ $R \rightarrow S$ $P \vee R$ $\therefore Q \vee S$	$((P \rightarrow Q) \wedge (R \rightarrow S) \wedge (P \vee R)) \rightarrow (Q \vee S)$	Constructive Dilemma

Modus Ponens (which roughly translated from Latin means "method of affirming") is probably the most familiar use of an implication in any argument. An implication and the antecedent of the implication are both assumed to be true and so the consequent must be true as well. **Modus Tollens** (roughly "method of denying") is the next most common argument with an implication. An implication is assumed to be true as is the negation of the consequent thus the antecedent must be false. We may use **Addition** to add any statement, via disjunction, to a statement already assumed to be true. **Simplification** allows us to manipulate a single part of a conjunctive proposition. **Conjunction** lets us conjoin any statements already assumed to be true as this does not affect their truth value in subsequent deductions.

Syllogisms are short valid arguments, like the rules of inference. There is some overlap between rules of inference and syllogisms but syllogisms traditionally only include arguments with two premises (a major premise and a minor premise) and a conclusion, and syllogisms also include certain arguments which are valid but require the use of quantified predicates. **Disjunctive Syllogism** is a valid argument based around a disjunctive proposition and the negation of one of the simple propositions in the disjunction. **Hypothetical Syllogism** allows us to essentially shorten a chain of implications into a single implication, it derives the "hypothetical" in its name from the conclusion which is an implication rather than a simple proposition, conjunction or disjunction.

The **Constructive Dilemma** can be used to leverage multiple implications and a disjunction in a similar manner to Modus Ponens. Note that the "Dilemma" in "Constructive Dilemma" has a different meaning here than what students are usually familiar with. The usual meaning is a choice between two undesirable outcomes. In logic it means two conditions imply the same conclusion and there is no value judgment about how desirable the conclusion may or may not be.

The rules of inference, given in Table 3.1, are of extreme importance because they greatly simplify the task of demonstrating that an argument form is valid. Recall that up until this point, our only methods of demonstrating that an argument form is valid is to either construct a truth table to demonstrate a tautology or manipulate a Boolean algebraic formula (such as $\neg(A \wedge B) \vee (A \vee B)$) until it reduced to a known tautology via one of the identity or complement laws.

Rules of inference let us demonstrate that an argument form is valid by starting with our premises and applying a series of valid argument forms to those premises. Whatever propositions we arrive at after applying a rule of inference we can safely conclude is guaranteed to be true, when the premises are all true. Thus, if we can eventually arrive at the desired conclusion using only rules of inference (or the laws of the algebra of propositions) we can safely state that the conclusion is guaranteed to be true when all the premises are true, and hence the argument form is valid.

We can find new argument forms by using the valid argument forms from Table 3.1.

Example 3.10
We start with an attempt to show that the argument form:
$A \rightarrow B$
$C \rightarrow D$
$\neg B$
$A \vee C$
$\therefore D$
is a valid argument.

First we number the premises. Then reference the numbered premises and the name of the rule of inference to draw a (temporary conclusion) which we also assign a new number to. This newly numbered temporary conclusion becomes a potential premise for further deduction.

1. $A \rightarrow B$ (Premise)
2. $C \rightarrow D$ (Premise)
3. $\neg B$ (Premise)
4. $A \vee C$ (Premise)
5. $\neg A$ (by Modus Tollens on 1 and 3)
6. C (by Disjunctive Syllogism on 4 and 5)
7. $\therefore D$ (by Modus Ponens on 2 and 6)

We could have also used:
5. $B \vee D$ (by Constructive Dilemma on 1,2 and 4)
6. $\therefore D$ (by Disjunctive Syllogism on 3 and 5)

There is not only one way of arriving at the conclusion using rules of inference. Keep in mind that if we had to compute the truth table for this argument it would have 32 columns for all the different combinations of T and F for the 4 variables A, B, C, and D. This is a much simpler method, and is the reason why rules of inference are the preferred method, of establishing that a complicated argument form is valid.

3.4   ADDITIONAL EXAMPLES

Example 3.11
Let P, Q, R be predicates.
Write the following propositions in symbolic notation:

1. For every x, If $P(x)$ and $R(x)$ then $Q(x)$.

2. There is an example where there is an x that satisfies $P(x) \wedge Q(x) \rightarrow R(x)$

3. It's not the case that whenever $P(x)$ implies $Q(x)$ then $P(x)$ also implies $R(x)$.

4. There exists an x such that $P(x) \vee Q(x)$ implies $R(x)$.

Solution:

1. $\forall x, P(x) \wedge R(x) \rightarrow Q(x)$

2. $\exists x, P(x) \wedge Q(x) \rightarrow R(x)$

3. $\neg(\forall x, (P(x) \rightarrow Q(x)) \rightarrow (P(x) \rightarrow R(x)))$

4. $\exists x, P(x) \vee Q(x) \rightarrow R(x)$

In the following examples we will be demonstrating arguments are valid using the rules of inference (along with the laws of the algebra of propositions) or we will be showing that an argument is invalid by providing a counterexample.

Example 3.12
$P \vee Q \rightarrow R$
$\neg R$
$\therefore \neg P \vee \neg Q$

Solution:

1. $P \vee Q \rightarrow R$ (Premise)
2. $\neg R$ (Premise)
3. $\neg(P \vee Q)$ (by Modus Tollens on 1 and 2)
4. $\neg P \wedge \neg Q$ (by De Morgan's Law on 3)
5. $\neg P$ (by Simplification on 4)
6. $\therefore \neg P \vee \neg Q$ (by Addition on 5)

Example 3.13

$A \rightarrow B$

$C \rightarrow A$

$C \rightarrow B$

$\therefore B \rightarrow A \wedge C$

Solution:

The argument is not valid. To see this, note that if $A = F$, $B = T$, and $C = F$ then all of the premises are true, but $B \rightarrow (A \wedge C)$ is false since B is true but $A \wedge C$ is false.

Example 3.14

$A \wedge B \wedge E$

$A \rightarrow C \vee D$

$C \rightarrow \neg E$

$D \rightarrow G$

$\therefore G \vee B$

Solution:

1. $A \wedge B \wedge E$ (Premise)
2. $A \rightarrow C \vee D$ (Premise)
3. $C \rightarrow \neg E$ (Premise)
4. $D \rightarrow G$ (Premise)

5. A (by Simplification on 1)
6. $C \vee D$ (by Modus Ponens on 2 and 5)
7. $G \vee \neg E$ (by Constructive Dilemma on 3, 4, and 6)
8. E (by Simplification on 1)
9. $\neg\neg E$ (by law of double negation on 8)
10. G (by Disjunctive Syllogism on 7 and 9)
11. $G \vee B$ (by Addition on 10)

3.5 PROBLEMS

Problem 3.15

Determine the universe of discourse for the following statements:

1. Student X has taken an advanced Calculus course.

2. For every even number x there is an odd number $x + 1$.

3. There is a function f which has a horizontal asymptote at 1.

4. For all $x > 2$, the function $f(x) = 5^x - 100$ is a positive number.

5. If s is the serial number of a registered piece of our software then $2^s - 1$ is a prime number.

Problem 3.16
Let A, B, C, and D be predicates.
Write the following propositions in symbolic notation:

1. For every x, $A(x)$ and $B(x)$ necessarily implies $C(x)$.

2. There is an x so that for every y, $A(x) \wedge B(y)$ means that $C(x) \vee D(y)$.

3. Whenever it's the case that there is some x such that $A(x)$ then $B(x)$.

4. It is not the case that there is an x where $A(x)$ and $B(x)$.

Problem 3.17
Determine the negation of the following propositions (and simplify the answer as much as possible):

1. $\forall x,\ P(x) \rightarrow Q(x) \wedge R(x)$

2. $\exists x,\ P(x) \wedge Q(x)$

3. $\forall x, \exists y,\ P(x) \wedge Q(y) \rightarrow A(x) \vee B(y)$

4. $\exists x,\ P(x) \wedge \exists y Q(y)$

For the following problems, if the argument is valid demonstrate it is valid using the rules of inference, like in *Example 3.12* **and** *Example 3.14*. **If the argument is not valid demonstrate it is not with a truth value assignment to the simple propositions which demonstrates that the premises can be true while the conclusion is still false, like in** *Example 3.13*.

Problem 3.18
It was raining outside or the wind was blowing. If the wind was blowing then I could not hear the crickets. It was raining outside. Therefore, I could hear the crickets.

Problem 3.19
If you sent me an email yesterday I would have written the reference letter before going to bed. If

you didn't send me an email yesterday I would go to sleep early. I went to sleep early. Therefore, you didn't send me an email yesterday.

Problem 3.20

$A \to B$

$B \to A$

$\therefore A \wedge B$

Problem 3.21

$A \to B \vee C$

$C \to D$

$\neg D \wedge A$

$\therefore B$

Problem 3.22

$P \wedge Q \to S$

$S \vee R \to S$

R

$\therefore \neg P$

Problem 3.23

$A \vee B \to C$

$C \to D \vee E$

$\neg E$

$\therefore \neg A \wedge \neg B$

Problem 3.24

$A \to B$

$B \to C$

$C \to A$

$\neg B$

$\therefore \neg A \wedge \neg C$

Problem 3.25

$A \vee B$

$C \to B$

$\neg C$

$\therefore A$

Problem 3.26

$A \vee B$

$B \rightarrow C$

$A \rightarrow D$

$\neg D$

$C \rightarrow A$

$\therefore A \wedge \neg C$

CHAPTER 4

Mathematical Proofs

4.1 SOME BACKGROUND

In the previous chapter we have seen what a valid argument is, and examined rules of inference. This chapter is concerned with mathematical proofs. **Mathematical proofs** are logical arguments that demonstrate a mathematical claim. There are several major categories of proofs. This chapter introduces some of the most important and commonly used ones. We shall start with a quick example proof.

Example 4.1
Claim: There are an infinite number of positive whole numbers.

Proof:

If there are not an infinite number of positive whole numbers then there must be a finite number n of whole numbers. Examine the list $1, 2, 3, \ldots, n$ of positive whole numbers. The list contains n whole numbers and so must therefore contain all positive whole numbers. We note, however, that the number $n + 1$ fails to appear on the list. This means the list is not complete. We have thus demonstrated that any finite list of integers is incomplete and deduce the list of all positive whole numbers is infinite. ☐

This is an informal proof, but it is a proof. In particular, it *states a claim which will be proven* then proceeds to demonstrate, by means of careful, precise language and deduction that the claim is true. The proof is an example of a proof technique called "proof by contradiction," where the opposite of what we are attempting to prove is assumed and used to derive a logical contradiction.

Unlike a valid argument form most proofs do not actually explicitly state every single assumption (premise) that is being used. For example, it is not explicitly stated that the two categories "finite" and "infinite" are mutually exclusive (no list or collection can be both finite and infinite) and exhaustive (there is no third option), nor is it stated that for any whole number n one may always construct a larger whole number $n + 1$. We instead proceed in a manner that is much closer to the use of the rules of inference from the previous chapter, each leap in logic is justified. For this to be useful or a "good proof" we do however, need to clarify the steps in logic to an extent that another reader can follow the chain of reasoning.

One of the requirements of a mathematical proof is that we have precise definitions from which to deductively reason. Often times we take these definitions for granted, like the definitions of "finite" and "infinite" in the previous example. However in mathematics one must always be able to have an explicit and precise definition available.

Take, for example, the idea of **even** and **odd** integers. If you are reading this text than you have obviously already encountered even and odd integers. If pressed you may reply that an even integer is an integer that is divisible by 2, and an odd one is not. For everyday language this is perfectly acceptable, however it does require one to also have a working definition of "divisible," because the entire definition is based on having or not having the property "divisible by 2." We may explicitly state the definition of "even number" by slightly modifying the definition to avoid the use of the word divisible.

Definition 4.2   An integer m is **even** if there exists an integer k such that $m = 2k$.

The definition of an odd number is slightly more complicated. We could define an odd number as an integer which is not an even number. This "negative definition" (a definition based on not having a property) may be equivalent to our usual notion of an "odd number" but there is one key problem with it. We are not particularly capable of explicitly representing an "odd number." We know that an odd number cannot look like $2k$ for any integer k but there is no explicit way (from the definition alone) to actually represent an odd number in a useful way. We proceed with a slightly modified definition of "odd number."

Definition 4.3   An integer m is **odd** if there exists an integer k such that $m = 2k + 1$.

This definition allows us to represent an odd number as $2k + 1$ for some k. What it does not do is explicitly state that a number cannot be both even and odd. Thus, if pressed we need to be able to prove that claim using only our definitions.

4.2   PROOF BY CONTRADICTION

We shall start with an example.

Example 4.4
Claim: Working in the domain of integers, an integer cannot be both even and odd.

Proof:

Assume the claim is false and that some integer n is both even and odd. Then there exist integers m and k such that $n = 2k$ and $n = 2m + 1$. Thus, $2k = 2m + 1$ and so $2(k - m) = 1$. Since k and m are integers $(k - m)$ is an integer so $2(k - m)$ is an even integer. However $2(k - m) = 1$. Thus, 1 is an even number. Since 1 is positive and an even number it must be a positive multiple of 2 and thus greater than 1. This is a contradiction since 1 cannot be greater than itself. Therefore, n cannot be both even and odd. $\qquad\qquad\qquad\qquad\qquad\qquad$ □

Proof by contradiction works as follows. We start with an implication $P \to Q$ which we want to prove. What does this mean? We want to show that the implication is a tautology. One way to do that is to show that $\neg(P \to Q)$ always has a truth value of false (a contradiction). Note that

$$\neg(P \to Q) \equiv \neg(\neg P \vee Q) \equiv \neg\neg P \wedge \neg Q \equiv P \wedge \neg Q$$

If we can show that $P \wedge \neg Q$ is a contradiction then we have shown that $P \to Q$ is a tautology. □

Example 4.5
Claim: The sum of two odd numbers is an even number.

Proof:

Assume, by way of contradiction, that a and b are odd numbers, and $a + b$ is not an even number. Since $a + b$ is not an even number, it is odd and can be written as $a + b = 2k + 1$ for some integer k. Since a and b are odd numbers, it follows that there are integers m and n such that $a = 2m + 1$ and $b = 2n + 1$. Thus, $2k + 1 = (2m + 1) + (2n + 1)$, therefore $2k + 1 = 2(m + n + 1)$ and hence $1 = 2(m + n + 1 - k)$ which implies that 1 is an even number. However, 1 is an odd number, since $1 = 2(0) + 1$. This means that 1 is both even and odd. This is a contradiction. Thus, if a and b are odd numbers, then their sum $a + b$ must be an even number.

The implication $P \to Q$ was not explicit in either of the previous two examples. In the first it can be unpacked as "If x is an integer then it is not the case that x is both even and odd," the second may be written as "If a and b are odd numbers then $a + b$ is an even number." This "hidden implication" is actually fairly common when everyday language is used to express mathematical ideas, again it is important that when pressed one can unravel such an implication into a more explicit "If . . . then" form.

Proof by contradiction rests on the fact that after the initial assumption of the negation of the claim that we can logically arrive at a contradiction (usually of the form $P \wedge \neg P$ for some proposition P). In the examples the contradictions were that 1 is greater than itself and that 1 is both an even and an odd number.

4.3   DIRECT PROOF

Direct proofs are also sometimes called **constructive proofs** or **proof by construction**. A **direct proof** is simply a series of deductions that lead from the antecedent of the implication to the consequent. That is all. Perhaps there are some tricky calculations, or a jumble of algebra to unscramble, but in a direct proof you assume all of the premises in the antecedent of your implication are true then demonstrate in a stepwise fashion that the conclusion must also be true. Generally this is done by carefully unpacking definitions.

Example 4.6
Claim: The sum of two even numbers is an even number.

Proof:

Let m and n be any two even numbers. Then since m is even there exists an integer k such that $m = 2k$, likewise there exists an integer j such that $n = 2j$. Note that $m + n = (2k) + (2j)$ thus, $m + n = 2(k + j)$. Since $m + n$ is a multiple of 2 it is an even number.

Let us examine the proof. We first unpack the claim as an implication, "the sum of two even numbers is an even number" becomes "if m and n are even numbers then their sum $m + n$ is an even number." We assume the initial assumption (that we have two even numbers). The definitions are unpacked. The definitions are applied to the initial assumption. This leads to some algebra or calculation or other logically acceptable manipulation until it becomes clear that the conclusion is satisfied. We then for good measure specify that the conclusion is satisfied. This is the basic idea behind (most) direct proofs.

Example 4.7
Claim: Using a 5-card poker hand from a standard pack of 52 cards. There are 624 four-of-a-kind poker hands.

Proof:

A poker hand is four-of-a-kind if 4 of the cards have the same value and the remaining card has a different value. We will count the number of four-of-a-kind hands by using independent choices to construct an arbitrary four-of-a-kind hand. There are 13 ways to pick the value for the four-of-a-kind. There is only 1 way to pick which suits appear in the four-of-a-kind, since all suits will need to be in the hand. There are 12 ways to pick the value of the other card and 4 ways to pick the suit for that card. This means there are $13(1)(12)(4) = 624$ different four-of-a-kind poker hands.

It may seem based on what we've stated that direct proofs are almost trivial. This is not necessarily the case. Sometimes some very clever ideas are needed to start at an appropriate point or to progress from one step to the next. Mathematics is more of an art than a science, and while there are common steps most direct proofs take, it is often not clear what the appropriate next step should be.

4.4   PROOF BY CONTRAPOSITION

Proof by contraposition works by switching an implication $P \to Q$ with it's contrapositive $\neg Q \to \neg P$, and then proceeding with a direct proof of the contrapositive. Since both implications are logically equivalent establishing the validity of one implication establishes the validity of the other.

Example 4.8
Claim: Let x and y be integers. If x is even and y is odd then $x + y$ is odd.

Proof:

The contrapositive of the claim is the claim: If $x + y$ is not odd, then it is not the case that, x is even and y is odd. Assume, by way of contraposition, that $x + y$ is not odd. Then $x + y$ is even and may be written as $x + y = 2k$ for some k. Thus, $x = 2k - y$. If y is odd then $y = 2m + 1$ for some m, and hence $x = 2k - 2m - 1 = 2(k - m - 1) + 1$ so x would be odd. So x is odd or y is even. Therefore, x is not even or y is not odd. Thus, by contraposition, if x is even and y is odd then $x + y$ is odd.

In the previous example it was not particularly easier to work with the contrapositive than the original implication but in some instances it may be. It is important that one does not confuse the contrapositive of the claim which is to be proven with the converse of the claim, as they are not logically equivalent.

The next example demonstrates a case where the contrapositive is easier to prove than the original implication.

Example 4.9
Claim: Let n be an integer. If n^2 is odd, then n is odd.

Proof:

Suppose that n is an even number. Then $n = 2k$ for some integer k. This means that $n^2 = (2k)(2k) = 2(2k^2)$ and hence n^2 is even. Since an integer cannot be both even and odd, this means that if n is not odd, then n^2 is not odd. By contraposition, this means that if n^2 is odd, then n is odd.

4.5   PROOF CONVENTIONS

In mathematics there are certain conventions for proofs that have been adopted. They do not actually add anything to the argument, other than brevity, clarity, or a certain elegance.

Nevertheless these are conventions, and students in mathematics should become familiar with them because they show up all over the place.

Mathematics is (almost always) written in the first person plural ("we, our"), never the first person singular ("I, my"). On some occasions it is written in the third person ("one") but this is generally considered overly formal and too passive a voice. The sciences generally use the third person, and it is because of this that Mathematics is occasionally written in this way, either because the author is so familiar with writing in the third person or because there are a few mathematical proofs in a science text. Mathematics is also generally written in the present tense ("We see that ...") and ("It follows that ...").

The universal and existential quantifiers used on propositional functions are used in proofs quite often. The symbol $\forall$ is used in place of the phrase "for all" and $\exists$ is used in place of the phrase "there exists." The phrase "such that" which frequently follows the existential quantifier is often replaced with "s.t." as an abbreviation.

Example 4.10
"For any integer k there exists an integer m such that $m > k$." may be written as "$\forall$ integers k, $\exists$ an integer m s.t. $m > k$."

The symbol $\Rightarrow$ is sometimes used in place of "implies" (and $\Leftrightarrow$ is used for if and only if) but is (generally) understood to mean that the writer of a proof is asserting the implication to be true, rather than merely making reference to the implication in a truth neutral manner. The symbol $\Rightarrow$ denotes a **logical consequence** (or **logical implication**) meaning the author is claiming that what follows is a logical consequence of what preceded it. The author is asserting that *material implication* is true. To make matters slightly more confusing sometimes $\Rightarrow$ is used to indicate that a chain of reasoning follows (is implied) by a previous chain of reasoning. Often the symbol $\Rightarrow$ is used to indicate that some algebra or calculation is implied, such as when one requires several steps to explain a particularly complicated manipulation of an equation. The symbol $\Rightarrow$ is read as "therefore" or "which implies" but is usually reserved for situations where everyday language such as "thus" or "it follows" would be used rather than an "if ... then."

The symbol $\Rightarrow\Leftarrow$ is used to denote that the chain of reasoning has arrived at a contradiction. The symbol $\therefore$ is used to abbreviate "therefore," particularly before a conclusion.

When ending a proof, the convention is to use either $Q.E.D.$ or the symbol $\square$ to indicate that the proof is finished. The $Q.E.D.$ is an abbreviation for the Latin phrase "quod erat demonstrandum" ("that which was to be demonstrated"), the $\square$ symbol is an alternative that is more common in modern usage. In either case, indicating that the proof is finished is considered good form. This is not just because it demonstrates that a student or novice to proofs actually knows when their proof is finished, but also allows for an easy way to visually distinguish portions of a text which are proofs, in case the reader wants to skip them or read them very carefully. Generally, the language in a proof is very precise and needs to be considered carefully if the reader is reading critically.

Occasionally, a proof will use the phrase "without loss of generality" or the equivalent abbreviation **WOLOG**. This is meant to indicate that a portion of the proof will be presented with an abstraction that requires an arbitrary choice or perspective, perhaps to a special case, but that the arbitrary choice does not actually affect the argument itself. The same steps will work with only minor adaptation for an alternate choice.

Example 4.11
Claim: Suppose there is a hand of five playing cards, with the cards having the standard suits of hearts, diamonds, spades, and clubs. Then there are at least two cards with the same suit.

Proof:

Assume WOLOG that the suit which most frequently occurs in the hand is clubs. Now assume by way of contradiction that there are five cards in the hand and that no suit occurs more than once. If there is only one club then since no suit occurs more frequently than a club there is at most one heart, one diamond, and one spade. However this accounts for only four cards, the fifth card cannot have a standard suit. This is a contradiction, therefore at least two cards in a hand of five have the same suit. □

The arbitrary choice in the previous example was to have clubs be the most frequent suit, but the same argument used in the proof holds if, for example, hearts was chosen instead of clubs.

In mathematics the claims that are proven are usually divided into one of three types. A **theorem** is a mathematical proposition which can be shown to be true. By this we mean that it's a tautology in the form of a valid argument form. A **lemma** is a theorem which is used to prove other theorems. A **corollary** is a theorem which follows directly from a theorem that has already been proved. For example, previously *an integer cannot be both even and odd* was used to demonstrate that *the sum of two odd numbers is an even number*. In this case *an integer cannot be both even and odd* is a lemma and *the sum of two numbers is an even number* is a corollary of the lemma.

In practice, however, the terms *lemma, theorem* and *corollary* are used slightly differently. A *theorem* is usually reserved for some major result, with wide-ranging applications. A *lemma* is a claim that is used to prove these important theorems, while a *corollary* is reserved for the direct results that follow from the important theorems. For minor results the terms **claim** or

proposition are usually used, although they may also be used for important results which do not quite merit the term theorem.

We will now revisit a couple earlier proofs with the standard conventions.

Example 4.12
Lemma: An integer cannot be both even and odd.

Proof:

Assume, by way of contradiction, that some integer n is both even and odd. Then $\exists$ integers m and k s.t. $n = 2k$ and $n = 2m + 1$.

$\Rightarrow 2k = 2m + 1 \Rightarrow 2(k - m) = 1$

Since k and m are integers $(k - m)$ is an integer so $2(k - m)$ is an even integer. However, $2(k - m) = 1.$ $\Rightarrow$ 1 is an even number.

Since 1 is positive and an even number it must be a positive multiple of 2 and thus greater than 1. However, 1 cannot be greater than itself. $\Rightarrow \Leftarrow$

$\therefore n$ cannot be both even and odd.   □

Example 4.13
Claim: The sum of two even numbers is an even number.

Proof:

Let m and n be any two even numbers.

$\Rightarrow \exists$ integers k and j such that $m = 2k$ and $n = 2j$.

We note that $m + n = (2k) + (2j) \Rightarrow m + n = 2(k + j)$.

Since $m + n$ is a multiple of 2 it is an even number.   □

Note that the $\Rightarrow$ symbols highlight the logical steps in the proofs. This allows the reader to grasp the essential temporary conclusions needed as further premises in the proofs.

4.6   COUNTEREXAMPLES AND DISPROVING A CLAIM

It is just as important to be able to deduce when a claim is false as when it is true. When we are talking about mathematical claims we are talking about tautologies, statements which are true regardless of the truth of the individual premises. Such claims are false when there is even one instance of the premises all being true while the conclusion is false. If one such instance can be found it is called a **counterexample**. These are **witnesses** which demonstrate that a claim is not always true.

Example 4.14
Claim: Let a and b be any two distinct integers; then the sum $a + b$ is a third distinct integer.

Counterexample:

Consider the case $a = 0$ and $b = 1$. Note that $a \neq b$ and $a + b = b$ since $0 + 1 = 1$.   □

4.7   CORRECT PROOFS AND THE HUMAN ELEMENT

The correctness of a proof is theoretically an objective, logically testable quality. The fact that the construction of proofs, as well as the reading and interpretation of proofs, happen in the human brain make this slightly less than true. In fact, the correctness of a proof is a consensus social judgment of the mathematical community. Look up the controversies surrounding the four-color theorem to get an idea about the kind of doubt a proof can engender.

Objectively correct proofs exist. They are usually incredibly long, very hard to follow, composed of small and unintuitive steps, and fail almost completely to convey understanding of the mathematical fact being demonstrated by the proof. A proof of this type is a *Tour-de-force* of little practical utility to those who are not hard-core mathematicians. Alfred North Whitehead and Bertrand Russell produced many proofs of this type in a three volume work entitled *Principia Mathematica*. In the *Principia Mathematica*, published in 1910–1913, the following appears on page 379 of the first edition: "From this proposition it will follow, when arithmetical addition has been defined, that 1+1=2." Yikes!

If things look a little bleak at this point fear not. Let us instead remember the point of a mathematical proof or argument in the first place. We are attempting to demonstrate that a conclusion follows logically from some initial set up. The word *demonstrate* is the key, a mathematical proof does not somehow bestow "truthness" on the claim. The claim was always true. The claim (including any necessary hidden premises) is and always was a tautology. All a good mathematical proof does is convince the reader that the claim is valid. This means that *the currency of proof is intuition*. If the person writing a proof and the person reading that proof have the same intuition about the mathematical structures in question then the proof may use that intuition to become shorter, easier to understand, and more beautiful. From this you can see that one of the great goals of mathematical training is common intuition. A good proof should help develop the intuition of the person reading it.

For all that we have been lauding intuition as the dragon-slayer of the immense, incomprehensible proof, we still want to stay as close to stepwise logical deductions as we can, using intuition only sparingly. Intuition has a power to mislead that is every bit as great as its ability to simplify and clarify. An incorrect intuition can sometimes get you completely backward and so you should maintain a watchful skepticism about your own intuition as it develops.

Once you have an idea for a proof the next step is to subject it to the eyes of your imagination. Think hard about what can go wrong. Does the proof work when $n = 0$? Are there any weird cases of the problem that you must separate out? Can the expression under the radical become negative? Use your imagination to question your proof idea. A proof you *cannot* demolish has a better chance of being correct.

If intuition is the currency of proof, then its wellsprings are precedent and imagination. Precedent because many correct proofs are constructed from other correct proofs by adapting the logic to new uses. Modus ponens, for example, probably arose several times as a proof of a single fact. Only later did some genius notice the commonality of those proofs and so discovered the general rule. Imagination is a wellspring of proof because it is the only tool available for reaching into an infinitude of possibilities and pulling out a workable approach. Precedent is more likely to work, so study all the proofs in your text and learn them. Imagination is more likely to impress your instructor or get you out of a tight corner. Imagination is also appallingly hard to practice. It just happens sometimes. On the other hand, you can develop your ability to use what your imagination hands you: "Chance favors only the prepared mind" - *Louis Pasteur [1854]*.

4.8   ADDITIONAL EXAMPLES

Example 4.15
Prove that the sum of two even numbers is an even number using a proof by contraposition.

Solution:

First we need to rewrite the claim as a conditional: If x and y are even numbers then $x + y$ is an even number. The contrapositive is: If $x + y$ is not an even number then x is not an even number or y is not an even number. We will use the fact that integers are either even or odd to simply the language to: If $x + y$ is an odd number then x or y is an odd number.

Proof (by Contraposition):

Let x and y be integers. Assume, by way of contraposition, that $x + y$ is an odd number. Then $\exists$ an integer m s.t. $x + y = 2m + 1$. So $x = 2m - y + 1$. If y is even then $\exists$ an integer n s.t. $y = 2n$, in this case $x = 2(m - n) + 1$ and since $m - n$ is an integer, x is an odd number. If y is not even then it is odd. So either x is an odd number or y is an odd number. Therefore we have that if $x + y$ is an odd number then x or y is an odd number. By contraposition we have that, if x and y are both even numbers, then $x + y$ is an even number. $\square$

Note: Obviously the proof by contraposition here is more clumsy than the direct proof, but it's worth practicing proof by contraposition now and again. Sometimes a proof is much easier by contraposition. Note that in the last sentence of the proof we included the original implication form of the claim in the proof toward the end. This is a vital step, it acts as a sanity check to ensure that the claim in plain English was translated into a conditional in an appropriate manner. It's also simply a nice thing to do for the reader of a proof.

Example 4.16

Prove that if a, b, and c are all integers and that $a + b = b + c$ and $b + c = a + c$ then a, b, and c are all the same integer.

Proof:

Let a, b, and c be integers. Assume (i) $a + b = b + c$ and (ii) $b + c = a + c$. Consider the expression $a + b + c$, since any expression is equal to itself we may deduce the following via substitution:

$$a + b + c = a + (a + c) = 2a + c$$

Subtracting a and c from both sides yields $b = a$. Similarly, we may deduce that:

$$a + b + c = (b + c) + c = b + 2c \Rightarrow a = c$$

Thus, $a = c$ and $a = b$ so $b = c \Rightarrow a,b$ and c are all the same number. $\square$

Note: The preceding proof is an example of when it's appropriate to use the symbol $\Rightarrow$. When it's appropriate is a bit subjective, but in general it's appropriate if it breaks the argument into bite-sized pieces which should be considered separately, such as temporary conclusions which will be used later. The proof is also an example of when it's appropriate to actually mention what is going on in a calculation. If the calculations were part of a much larger problem it likely wouldn't be necessary to mention exactly what was subtracted from both sides but since the entire argument in this proof rests on a line of *calculations it's worth explaining what is actually going on.*

Example 4.17   Prove that the product of two consecutive integers is an even number.

Proof:

Let x and $x + 1$ be two consecutive integers.

Case 1: x is an odd number.
If x is odd then there exists an integer m such that $x = 2m + 1$, thus:

$$x(x + 1) = (2m + 1)(2m + 1 + 1) = (2m + 1)(2m + 2) = 4m^2 + 6m + 2$$
$$= 2(2m^2 + 3m + 1)$$

since m is an integer $2m^2 + 3m + 1$ is an integer and we have that $x(x + 1)$ is an even number.

Case 2: x is an even number.
If x is even then there exists an integer n such that $x = 2n$, thus:

$$x(x + 1) = 2n(2n + 1) = 2(n(2n + 1))$$

since n is an integer $n(2n + 1)$ is an integer and hence $x(x + 1)$ is an even number.
Therefore, in all cases, $x(x + 1)$ is an even number.   □

Note: The logic of the preceding argument rests on a constructive dilemma, we demonstrate that two conditional implications are a consequence of our initial assumption. Then we combine those conditionals with a tautology of the form $(P \lor \neg P)$ to get the constructive dilemma. Sometimes the tautology of the form $P \lor \neg P$ is implicit, like in the preceding proof where we used the fact that an integer is either even or odd without explicitly stating it. Proofs of this sort (or similar) are called proofs with cases, or sometimes **proof by exhaustion.** *If a teacher or someone else mentions that a particular problem is easier if you break it into cases then such a proof will likely use, a possibly modified form of, the constructive dilemma.*

4.9   PROBLEMS

Problem 4.18
Prove the following claim using a direct proof: *the product of two even numbers is an even number* (i.e., show that if x and y are even numbers then xy is an even number).

Problem 4.19
Prove the following claim using a proof by contradiction: *If the product of two integers x and y is odd then at least one of x or y is odd.*

Problem 4.20
Prove the following claim using a proof by contraposition: *the product of two odd numbers is an odd number*

Problem 4.21
Prove or disprove: If x and y are integers and $xy = x$, then $x = 1$ or $y = 1$.

Problem 4.22
Prove or disprove: Suppose $a + b$ and ab are both even numbers. Then a and b are both even numbers.

Problem 4.23
Presented below are four different "proofs" for the same claim. Read the proofs and answer the questions that follow.

Claim: The sum of two consecutive integers is an odd number.

Proof 1: Assume without loss of generality that a and $a + 1$ are two consecutive integers. Suppose, by way of contradiction, that $a + (a + 1)$ is even.
$\Rightarrow a + (a + 1) = 2k$ for some integer k.
$\Rightarrow 2a + 1 = 2k \Rightarrow a + (a + 1)$ is both even and odd. This is a contradiction.
$\therefore a + (a + 1)$ is odd.   □

Proof 2: Let a and b be consecutive integers. Then $a + b$ is odd.

Proof 3: Let a and $a + 1$ be consecutive integers.
If a is even then there exists k such that $a = 2k$. $\Rightarrow a + (a + 1) = 2k + (2k + 1) = 2(2k) + 1$ which is an odd number.
If a is odd then there exists k such that $a = 2k + 1$. $\Rightarrow a + (a + 1) = (2k + 1) + ((2k + 1) + 1) = 2(2k + 1) + 1$ which is an odd number.
$\therefore a + (a + 1)$ is odd. Q.E.D.

Proof 4: Let a and b be integers. Suppose, by way of contradiction, that $a + b$ is even.
Then, $\exists$ k,m such that $a = 2k$ and $b = 2m$.

$\Rightarrow b \neq a + 1$ and $a \neq b + 1 \therefore a$ and b are not consecutive integers.   □

Proof 5: Suppose $x + y$ is an even number. Then x is odd or x is even.
If x is odd then y has to be odd. If x is even then y has to be even. Either way x and y are both the same type, either even or odd. Therefore, x and y cannot be consecutive integers.   □

Answer the following questions:

1. The crux of the argument in Proof 3 rests on an unstated rule of inference. What rule is it?

2. What type of proof (direct, contradiction, contraposition) is Proof 4? Explain.

3. Which of these proofs are actually correct?

4. For the proof(s) which are incorrect where *exactly* is/are the error(s)?

5. Technically, Proof 2 is valid. Nevertheless, it is not a good proof. Under most circumstances it would be considered incorrect. What makes it a bad proof?

6. In Proof 5 the claim in each sentence of the argument is true. Does this mean that it's a good proof?

CHAPTER 5

Intuitive Set Theory

The material in this chapter is elementary. Elementary does not mean *simple* (though much of the material is fairly simple). Rather, elementary means that the material requires very little previous education to understand it. Elementary material can be quite challenging and some of the material in this chapter, if not exactly rocket science, will require that you adjust your point of view to understand it. The single most powerful technique in mathematics is to adjust your point of view until the problem you are trying to solve becomes simple.

5.1   SET THEORY

Set theory originated with Georg Cantor (1845–1918). During Cantor's mathematical investigations he realized a need for comparing the magnitude, or size, of various infinite sets of numbers. Not only did this lead to a way of comparing different infinities, and showing that, for instance, all the real numbers are a much larger infinite set than all of the integers, but this led to all sorts of innovations and ultimately toward providing a solid foundation for mathematics.

Before getting too far along, it's worth noting that we will be refering to **intuitive set theory** also known as **naive set theory** in this chapter. This is, more or less, the set theory Cantor originally proposed. Modern set theory can at times look a bit different but an understanding of naive set theory is a necessary first step.

So what is a set? Cantor's definition is that a **set** is *any collection of definite, distinguishable objects of the intellect which is conceived of as a whole.*

We shall use a slightly more modern definition.

Definition 5.1   A *set* is a well-defined collection of distinct objects.

This means that $\{1, 2, 3\}$ is a set but $\{1, 1, 3\}$ is not because 1 appears twice in the second collection. The second collection is called a **multiset**. Sets are often specified with curly brace notation. The set of even integers can be written:

$$\{2n : n \text{ is an integer}\}$$

The opening and closing curly braces denote a set, $2n$ specifies the members of the set, the colon says "such that" or "where" and everything following the colon are conditions that explain or refine the membership. All correct mathematics can be spoken in English. The set definition above is spoken "The set of twice n where n is an integer."

The only problem with this definition is that we do not yet have a formal definition of the integers. For now we'll say the **integers** are the set of whole numbers, both positive and negative (and including zero): $\{0, \pm 1, \pm 2, \pm 3, \ldots\}$. The standard mathematical symbol for integers is $\mathbb{Z}$ (the Z is from the german word for numbers, *zahlen*).

This explicit listing of the elements of a set is called the **tabular form** of a set. Specifying a set by means of a predicate that all elements of the set must satisfy is called **set builder notation**. These are the usual methods for specifying a set. There are rare cases when other means may be used to specify sets but these are not common even in abstract mathematics, the tabular form and set builder notation is usually just shortened to **set notation**.

We now introduce the operations used to manipulate sets, using the opportunity to practice set notation.

Definition 5.2   The **empty set** is a set containing no objects. It is written as a pair of curly braces with nothing inside $\{\}$ or by using the symbol $\emptyset$. The empty set is sometimes called the **null set**.

As we shall see, the empty set is a handy object. It is also quite strange. The set of all humans that weigh at least eight tons, for example, is the empty set. Sets whose definition contains a contradiction or impossibility are often empty.

Definition 5.3   The **set membership symbol** $\in$ is used to say that an object is a member of a set. It has a partner symbol $\notin$ which is used to say an object is not in a set.

Sets are usually denoted by uppercase letters such as $A, B, C \ldots$, while lowercase letters like a, b, and c are usually used to denote elements of a set. The set membership symbol is often used in defining operations that manipulate sets.

Definition 5.4   We say two sets are **equal** if and only if they have exactly the same members.

Example 5.5
If

$$S = \{1, 2, 3\}$$

then $3 \in S$ and $4 \notin S$. The set

$$T = \{2, 3, 1\}$$

is equal to S because they have the same members: 1, 2, and 3. While we list the members of a set in a "standard" order (if one is available) there is no requirement to do so and sets are indifferent to the order in which their members are listed.

Definition 5.6   The **cardinality** of a set is its *size*. For a finite set, the cardinality of a set is the number of members it contains. In symbolic notation the size of a set S is written $|S|$. We will deal with the idea of the cardinality of an infinite set later.

Example 5.7
For the set $S = \{1, 2, 3\}$ we show cardinality by writing $|S| = 3$.

We now move on to a number of **operations** on sets. You are already familiar with several operations on numbers such as addition, multiplication, and negation.

Definition 5.8   The **intersection** of two sets S and T is the collection of all objects that are in both sets. It is written $S \cap T$. Using set notation

$$S \cap T = \{x : (x \in S) \text{ and } (x \in T)\}$$

The word *and* in the above definition is an example of a Boolean or logical operation, its symbolic equivalent is $\wedge$. This lets us write the formal definition of intersection more compactly:

$$S \cap T = \{x : (x \in S) \wedge (x \in T)\}$$

Example 5.9
Suppose $S = \{1, 2, 3, 5\}$, $T = \{1, 3, 4, 5\}$, and $U = \{2, 3, 4, 5\}$.
Then:

$S \cap T = \{1, 3, 5\}$,

$S \cap U = \{2, 3, 5\}$, and

$T \cap U = \{3, 4, 5\}$

Definition 5.10   If A and B are sets and $A \cap B = \emptyset$ then we say that A and B are *disjoint*, or *disjoint sets*.

Definition 5.11   The **union** of two sets S and T is the collection of all objects that are in either set. It is written $S \cup T$. Using set notion

$$S \cup T = \{x : (x \in S) \text{ or } (x \in T)\}$$

The word *or* is another Boolean operation, its symbolic equivalent is $\vee$ which lets us re-write the definition of union as:

$$S \cup T = \{x : (x \in S) \vee (x \in T)\}$$

Example 5.12
Suppose $S = \{1, 2, 3\}$, $T = \{1, 3, 5\}$, and $U = \{2, 3, 4, 5\}$.
Then:

$S \cup T = \{1, 2, 3, 5\}$,

$S \cup U = \{1, 2, 3, 4, 5\}$, and

$T \cup U = \{1, 2, 3, 4, 5\}$

When performing set theoretic computations, you should declare the domain in which you are working. In set theory this is done by declaring a universal set.

Definition 5.13   The **universal set**, at least for a given collection of set theoretic computations, is the set of all possible objects.

If we declare our universal set to be the integers then $\{\frac{1}{2}, \frac{2}{3}\}$ is not a well-defined set because the objects used to define it are not members of the universal set. The symbols $\{\frac{1}{2}, \frac{2}{3}\}$ do define a set if a universal set that includes $\frac{1}{2}$ and $\frac{2}{3}$ is chosen. The problem arises from the fact that neither of these numbers are integers. The universal set is commonly written $\mathcal{U}$.

5.2   VENN DIAGRAMS

A Venn diagram is a way of depicting the relationship between sets. Each set is shown as a circle and circles overlap if the sets intersect.

Example 5.14
On the following page are Venn diagrams for the intersection and union of two sets. The shaded parts of the diagrams are the intersections and unions, respectively.

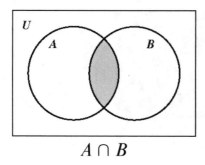 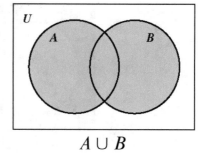

$$A \cap B \qquad\qquad A \cup B$$

Notice that the rectangle containing the diagram is labeled with a U representing the universal set.

Definition 5.15 The **complement** of a set S is the collection of objects in the universal set that are not in S. The complement is written S^c. In curly brace notation

$$S^c = \{x : (x \in \mathcal{U}) \wedge (x \notin S)\}$$

or more compactly as

$$S^c = \{x : x \notin S\}$$

however it should be apparent that the complement of a set always depends on which universal set is chosen.

There is also a Boolean symbol associated with the complementation operation: the *not* operation. The not operation reverses the truth value, turning true into false and false into true. The notation for not is $\neg$. There is not much savings in space as the definition of complement becomes

$$S^c = \{x : \neg(x \in S)\}$$

Example 5.16

1. Let the universal set be the integers. Then the complement of the even integers is the odd integers.

2. Let the universal set be $\{1, 2, 3, 4, 5\}$, then the complement of $S = \{1, 2, 3\}$ is $S^c = \{4, 5\}$ while the complement of $T = \{1, 3, 5\}$ is $T^c = \{2, 4\}$.

3. Let the universal set be the letters $\{a, e, i, o, u, y\}$. Then $\{y\}^c = \{a, e, i, o, u\}$.

The Venn diagram for A^c is

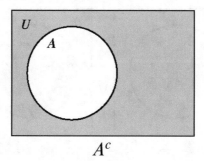

$$A^c$$

We now have enough set-theory operators to use them to define more operators quickly. We will continue to give English and symbolic definitions.

Definition 5.17   The **difference** of two sets S and T is the collection of objects in S that are not in T. The difference is written $S - T$. In curly brace notation

$$S - T = \{x : x \in (S \cap (T^c))\},$$

or alternately

$$S - T = \{x : (x \in S) \wedge (x \notin T)\}$$

5.3   OTHER OPERATIONS ON SETS

Notice how intersection and complementation can be used together to create the difference operation and that the definition can be rephrased to use Boolean operations. There is a set of rules that reduces the number of parenthesis required. These are called **operator precedence rules**.

1. Other things being equal, operations are performed left-to-right.

2. Operations between parenthesis are done first, starting with the innermost of nested parenthesis.

3. All complementations are computed next.

4. All intersections are done next.

5. All unions are performed next.

6. Tests of set membership and computations, equality or inequality are performed last.

The operator precedence rules correspond to the rules for the precedence of logical connectives. The symbol ¬ corresponds to set complement, ∧ to intersection, and ∨ to union. Special operations like the set difference or the symmetric difference, defined below, are not included in the precedence rules and thus always use parentheses.

Example 5.18
Since complementation is done before intersection the symbolic definition of the difference of sets can be rewritten:

$$S - T = \{x : x \in S \cap T^c\}$$

If we were to take the set operations

$$A \cup B \cap C^c$$

and put in the parenthesis we would get

$$(A \cup (B \cap (C^c)))$$

Definition 5.19   The **symmetric difference** of two sets S and T is the set of objects that are in one and only one of the sets. The symmetric difference is written $S\Delta T$. In set notation:

$$S\Delta T = \{(S - T) \cup (T - S)\}$$

Example 5.20
Let S be the set of non-negative multiples of 2 that are no more than 24. Let T be the non-negative multiples of three that are no more than 24. Then

$$S\Delta T = \{2, 3, 4, 8, 9, 10, 14, 15, 16, 20, 21, 22\}$$

Another way to think about this is that the symmetric difference of S and T is the set of numbers that are positive multiples of 2 or 3 (but not both) that are no more than 24.

Another important tool for working with sets is the ability to compare them. We have already defined what it means for two sets to be equal, and so by implication what it means for them to be unequal. We now define another comparator for sets.

Definition 5.21   For two sets S and T we say that S **is a subset of** T if each element of S is also an element of T. In formal notation $S \subseteq T$ if for all $x \in S$ we have $x \in T$.

If $S \subseteq T$ then we also say T contains S which can be written $T \supseteq S$. If $S \subseteq T$ and $S \neq T$ then we write $S \subset T$ and we say S is a **proper subset** of T.

Example 5.22

If $A = \{a, b, c\}$ then A has eight different subsets:

$$\emptyset \qquad \{a\} \qquad \{b\} \qquad \{c\}$$

$$\{a, b\} \quad \{a, c\} \quad \{b, c\} \quad \{a, b, c\}$$

Notice that $A \subseteq A$ and in fact each set is a subset of itself. The empty set $\emptyset$ is a subset of every set.

5.4 BASIC RESULTS

We are now ready to prove our first set theory claim, but first a quick reminder. If we say "A if and only if B" then we mean that either A and B are both true or they are both false in any given circumstance. For example: "an integer x is even if and only if it is a multiple of 2." The phrase "if and only if" is used to establish *logical equivalence*. Mathematically, "A if and only if B" is a way of stating that A and B are simply different ways of saying the same thing. The phrase "if and only if" is abbreviated iff and is represented symbolically as the double arrow $\Leftrightarrow$. Proving an iff statement is done by independently demonstrating that each proposition may be deduced from the other.

Proposition 5.23   Two sets are equal if and only if each is a subset of the other. In symbolic notation:

$$(A = B) \quad \Leftrightarrow \quad (A \subseteq B) \wedge (B \subseteq A)$$

Proof:

This is a direct proof. Let the two sets in question be A and B. Begin by assuming that $A = B$. We know that every set is a subset of itself so $A \subseteq A$. Since $A = B$ we may substitute into this expression on the left and obtain $B \subseteq A$. Similarly we may substitute on the right and obtain $A \subseteq B$. We have thus demonstrated that if $A = B$ then A and B are both subsets of each other, giving us the first half of the iff.

Assume now that $A \subseteq B$ and $B \subseteq A$. Then the definition of subset tells us that any element of A is an element of B. Similarly, any element of B is an element of A. This means that A and $B have the same elements which satisfies the definition of set equality. We deduce $A = B$ and we have the second half of the iff. □

Proposition 5.24   De Morgan's Laws
Suppose that S and T are sets. De Morgan's Laws state that

(i) $(S \cup T)^c = S^c \cap T^c$, and

(ii) $(S \cap T)^c = S^c \cup T^c$.

Proof:

Let $x \in (S \cup T)^c$; then x is not a member of S or T. Since x is not a member of S we see that $x \in S^c$. Similarly, $x \in T^c$. Since x is a member of both these sets we see that $x \in S^c \cap T^c$ and we see that $(S \cup T)^c \subseteq S^c \cap T^c$. Let $y \in S^c \cap T^c$. Then the definition of intersection tells us that $y \in S^c$ and $y \in T^c$. This in turn lets us deduce that y is not a member of $S \cup T$, since it is not in either set, and so we see that $y \in (S \cup T)^c$. This demonstrates that $S^c \cap T^c \subseteq (S \cup T)^c$. Applying Proposition 5.23 we get that $(S \cup T)^c = S^c \cap T^c$ and we have proven part (i). The proof of part (ii) is left as an exercise. □

5.5   ADDITIONAL EXAMPLES

Example 5.25
Prove that the statement $A \cup B = A \cap B$ is false.

Solution:

Let $A = \{1, 2\}$ and $B = \{3, 4\}$. Then $A \cap B = \emptyset$ while $A \cup B = \{1, 2, 3, 4\}$. The sets A and B form a counterexample to the statement. The sets $A = \{1, 2\}$ and $B = \{3, 4\}$ are witnesses to the fact that $A \cup B = A \cap B$ is not true in general.

Example 5.26
Suppose that we have the set $U = \{n : 0 \leq n \leq 100\}$ of whole numbers as our universal set. Let A be all multiples of 5 in U, let B be all multiples of 10 in U, and let $C = \{30, 31, 33, 35, 37, 40\}$. Describe the following sets either by listing them or with a careful English sentence:

(i) $A \cap B^c$,

(ii) $B \cap C$,

(iii) $A \cup B$,

(iv) $A^c \cap C \cup B^c$

Solution:

(i) All multiples of 5 which aren't multiples of 10 that are in U.

(ii) $\{30, 40\}$

(iii) All multiples of 5 in U. (Note: $B \subset A$)

(iv) All members of U which aren't multiples of 10. (Note: $A^c \cap C = \{31, 33, 37\} \subset B^c$

Example 5.27
Prove that $(A \cup B) \cap C = (A \cap C) \cup (B \cap C)$.

Proof:

First we will show that $(A \cup B) \cap C \subseteq (A \cap C) \cup (B \cap C)$.

Let $x \in ((A \cup B) \cap C)$, then $x \in (A \cup B)$ and $x \in C$. Therefore $x \in A$ or $x \in B$ and also $x \in C$. By the distributive law (for propositions) we have that $x \in A$ and $x \in C$ or $x \in B$ and $x \in C$, thus $x \in (A \cap C) \cup (B \cap C)$.

Similarly, by applying the distributive law (for propositions) in the opposite direction, it's clear that $(A \cap C) \cup (B \cap C) \subseteq (A \cup B) \cap C$. Thus, both sets are subsets of each other and the result follows.   □

Note: In the preceding proof we did not bother to go through the tedious process to demonstrate that the set on the right-hand side was also a subset of the set on the left-hand side. If we had it would have literally been an issue of writing the same lines over again but in the opposite order. There is nothing wrong with avoiding tedious portions of a proof like this, as long as it is obvious and nothing noteworthy is skipped. When it is or is not acceptable to skip over certain steps, is at times a subjective matter with whomever the argument is directed at (remember all proofs are arguments). The general rule is that it must be possible to explain the section in detail, and with minimal effort, should the need arise.

5.6   PROBLEMS

Problem 5.28
Which of the following are sets? Assume that a proper universal set has been chosen and answer by listing the names of the collections of objects that are sets. Warning: at least one of these items has an answer that, while likely, is not 100% certain.

(i) $A = \{2, 3, 5, 7, 11, 13, 19\}$

(ii) $B = \{a, e, i, o, u\}$

(iii) $C = \{\sqrt{x} : x < 0\}$

(iv) $D = \{1, 2, a, 5, b, q, 1, v\}$

(v) E is the list of first names of people in the 1972 phone book in Lawrence Kansas in the order they appear in the book. There were more than 35,000 people in Lawrence that year.

(vi) F is a list of the weight, to the nearest kilogram, of all people that were in Canada at any time in 2007.

(vii) G is a list of all weights, to the nearest kilogram, that at least one person in England had in 2019.

Problem 5.29
Suppose that we have the set $U = \{n : 0 \le n < 100\}$ of whole numbers as our universal set. Let P be the prime numbers in U, let E be the even numbers in U, and let $F = \{1, 2, 3, 5, 8, 13, 21, 34, 55, 89\}$. Describe the following sets either by listing them or with a careful English sentence.

(i) E^c,

(ii) $P \cap F$,

(iii) $P \cap E$,

(iv) $F \cap E \cup F \cap E^c$, and

(v) $F \cup F^c$.

Problem 5.30
Suppose that we take the universal set $\mathcal{U}$ to be the integers. Let S be the even integers, let T be the integers that can be obtained by tripling any one integer and adding one to it, and let V be the set of numbers that are whole multiples of both two and three.

(i) Write S, T, and V using symbolic notation.

(ii) Compute $S \cap T$, $S \cap V$, and $T \cap V$ and give symbolic representations that do not use the symbols S, T, or V on the right-hand side of the equals sign.

Problem 5.31

Compute the cardinality of the following sets. You may use other texts or the internet.

(i) Two digit positive odd integers.

(ii) Positive multiples of seven that are less than 100.

(iii) Planets orbiting the same star as the planet you are standing on that have moons. Assume that Pluto is not a planet.

(iv) Subsets of $\{0, 1\}$.

(v) Solutions of the equation $x^2 - 4x + 3 = 0$.

(vi) Subsets of $S = \{a, b, c, d\}$ with cardinality 2.

Problem 5.32

Find an example of an infinite set that has a finite complement, be sure to state the universal set.

Problem 5.33

Find an example of an infinite set that has an infinite complement, be sure to state the universal set.

Problem 5.34

Add parenthesis to each of the following expressions that enforce the operator precedence rules as in Example 5.18. Notice that the first three describe sets while the last returns a logical value (true or false).

(i) $A \cup B \cup C \cup D$

(ii) $A \cup B \cap C \cup D$

(iii) $A^c \cap B^c \cup C$

(iv) $A \cup B = A \cap C$

Problem 5.35

Examine the Venn diagram on the following page. Notice that every combination of sets has a unique number in common. Construct a similar collection of four sets.

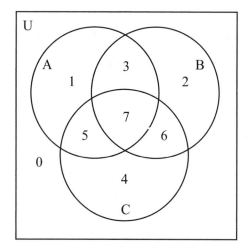

Problem 5.36
Read Problem 5.35. Can a system of sets of this sort be constructed for any number of sets? Explain your reasoning.

Problem 5.37
Give the Venn diagrams for the following sets.

(i) $A - B$ (ii) $B - A$ (iii) $A^c \cap B$ (iv) $A \triangle B$ (v) $(A \triangle B)^c$ (vi) $A^c \cup B^c$

Problem 5.38
The Fibonacci numbers are given by the sequence $0, 1, 1, 2, 3, 5, 8, \ldots$ in which the next term is obtained by adding the previous two. Suppose we take the universal set to be the set of non-negative integers. Let E be the set of even numbers, O be the set of odd numbers, and $F = \{0, 1, 2, 3, 5, 8, 13, 21, 34, 89, 144, \ldots\}$ be the set of Fibonacci numbers.

(i) Prove that the intersection of F with E and O are both infinite.

(ii) Make a Venn diagram for the sets E, F, and O, and explain why this is a mouse-related problem.

Problem 5.39
A binary operation $\odot$ is **commutative** if $x \odot y = y \odot x$. An example of a commutative operation is multiplication. Subtraction is non-commutative. Determine, with proof, if union, intersection, set difference, and symmetric difference are commutative.

Problem 5.40

An **identity** for an operation $\odot$ is an object i so that, for all objects x, $i \odot x = x \odot i = x$. Find, with proof, identities for the operations *set union* and *set intersection*.

Problem 5.41

Prove part (ii) of Proposition 5.24.

Problem 5.42

Prove that

$$A \cup (B \cup C) = (A \cup B) \cup C$$

Problem 5.43

Prove that

$$A \cap (B \cap C) = (A \cap B) \cap C$$

Problem 5.44

Prove that

$$A \triangle (B \triangle C) = (A \triangle B) \triangle C$$

Problem 5.45

Disprove that

$$A \triangle (B \cup C) = (A \triangle B) \cup C$$

Problem 5.46

Prove that

$$(A \cap B) \cup C = (A \cup C) \cap (B \cup C)$$

Problem 5.47

Consider the set $S = \{1, 2, 3, 4\}$. For each $k = 0, 1, \ldots, 4$, how many k element subsets does S have?

Problem 5.48
For finite sets S and T, prove

$$|S \cup T| = |S| + |T| - |S \cap T|$$

CHAPTER 6

Mathematical Induction

Mathematical induction is a proof technique useful for proving a certain type of mathematical assertion. The basic idea of induction is that we prove that a statement is true in one case and then also prove that if it is true in a given case it is true in the next case. This then permits the cases for which the statement is true to cascade from the initial true case, like knocking down a row of dominos.

Mathematical induction is the last of the major proof techniques we will cover. Using mathematical induction as a proof technique is a little different than the other techniques. We must be very careful when demonstrating that the truth of the next case is actually implied by the truth of the previous cases. First, we will give the reader a quick example of a proof by induction, before we demonstrate that it is in fact a valid proof technique.

We assume that the reader is familiar with the symbols $<$, $>$, $\leq$, and $\geq$. From this point on we will denote the set of integers by the symbol $\mathbb{Z}$. The non-negative integers are called the **natural numbers**. The symbol for the set of natural numbers is $\mathbb{N}$.

We will begin with a quick, informal example of mathematical induction.

Example 6.1
Claim: For all $n \in \mathbb{N}$, $n \leq 2^n$.

Proof:

Let $P(n)$ be the predicate, $n \leq 2^n$. We note that when $n = 0$, $2^0 = 1$. So in this case $P(0)$ is true.

Let us assume that the predicate $P(n)$ is true for some natural number k. Then $k \leq 2^k \Rightarrow (k+1) \leq 2^k + 1$.

As long as $k \geq 0$ then $2^k + 1 \leq 2^k + 2^k \leq 2(2^k) \Rightarrow 2^k + 1 \leq 2^{k+1}$. Thus, $(k+1) \leq 2^{k+1}$.

We've shown that whenever $P(k)$ is true then $P(k+1)$ is true, we've also shown that $P(0)$ is true. Therefore, $P(n)$ is true for all $n \geq 0$. $\square$

Now that we've seen an example, it's time to seriously consider whether mathematical induction is even a valid proof technique.

Any mathematical system rests on a foundation of axioms. **Axioms** are the foundational premises we need to assume are true in order to gain any traction. Axioms are different than hidden premises. If the reader of a proof or someone we are trying to convince fails to see why a hidden premise is true we need to be able to justify it. Axioms are never justified. Axioms are things that we simply assume to be true. Different systems of mathematics may take different axioms as a given, and consequently propositions which are demonstrably true in one system may not be in another. We will discuss axioms in greater depth later in the text, for now we will assume the truth of the following principle, adopting it as an axiom.

Definition 6.2   The Well-Ordering Principle
Every non-empty set of natural numbers contains a smallest element.

Since we are adopting the **Well-Ordering Principle** as an axiom we do not need to justify it, however it is worth noting that it is an axiom that agrees with the common sense of most people familiar with the natural numbers. An empty set does not contain a smallest member because it contains no members at all. As soon as we have a set of natural numbers with some members then we can order those members in the usual fashion. Having ordered them, one will be smallest.

The intuition agreeing with this claim depends strongly on the fact the integers are "whole numbers" spaced out in increments of one. To see why this is important consider the smallest positive distance, which is a real number rather than a whole number. If such a distance existed, we could cut it in half to obtain a smaller distance—the quantity contradicts its own existence. The Well-Ordering Principle can be used to prove the validity of induction.

Theorem 6.3   Mathematical Induction
Suppose that for every given natural number n, $P(n)$ is a proposition that is either true or false. If

(i) *$P(0)$ is true and,*

(ii) *when $P(n)$ is true so is $P(n + 1)$*

Then we may deduce that $P(n)$ is true for any natural number.

Proof:

Assume that (i) and (ii) are both true statements. Let S be the set of all natural numbers for which $P(n)$ is false. If S is empty then we are done, so assume that S is not empty.

Then, by the Well-Ordering Principle, S has a least member m. By (i) above $m \neq 0$ and so $m - 1$ is a natural number. Since m is the smallest member of S it follows that $P(m - 1)$ is true. But this means, by (ii) above, that $P(m)$ is true. We have a contradiction and so our assumption that $S \neq \emptyset$ must be wrong. We deduce S is empty and that as a consequence $P(n)$ is true for all $n \in \mathbb{N}$. □

The above proof is an example of *proof by contradiction*. We start by assuming the logical opposite of what we want to prove, in this case that there is some m for which $P(m)$ is false, and from that assumption we derive an impossibility. Thus, the validity of the proof by contradiction technique, along with the axiom of the Well-Ordering Principle demonstrates the validity of mathematical induction *as a proof technique*.

A nice problem on which to demonstrate mathematical induction is counting how many subsets a finite set has.

Proposition 6.4   A set S with n elements has 2^n subsets.

Proof:

First, we check that the proposition is true when $n = 0$. The empty set has exactly one subset: itself. Since $2^0 = 1$ the proposition is true for $n = 0$. We now assume the proposition is true for some n. Suppose that S is a set with $n + 1$ members and that $x \in S$. Then $S - \{x\}$ (the set difference of S and a set $\{x\}$ containing only x) is a set of n elements and so, by the assumption, has 2^n subsets. Now every subset of S either contains x or it fails to. Every subset of S that does not contain x is a subset of $S - \{x\}$ and so there are 2^n such subsets of S. Every subset of S that contains x may be obtained in exactly one way from one that does not by taking the union with $\{x\}$. This means that the number of subsets of S containing or failing to contain x are equal. This means there are 2^n subsets of S containing x. The total number of subsets of S is thus $2(2^n) = 2^{n+1}$. Thus, we have demonstrated that whenever the proposition is true for n it is also true for $n + 1$. It follows by mathematical induction that the proposition is true for all natural numbers. □

The set of all subsets of a given set is itself an important object and so has a name.

Definition 6.5   The set of all subsets of a set S is called the **power set** of S. The notation for the power set of S is $\mathcal{P}(S)$.

This definition permits us to rephrase Proposition 6.4 as follows: the power set of a set of n elements has size 2^n.

Theorem 6.3 lets us prove propositions that are true on the natural numbers, starting at zero. A small modification of induction can be used to prove statements that are true only for those $n \geq k$ for any integer k. All that is needed is to use induction on a proposition $Q(n-k)$ where $Q(n-k)$ is logically equivalent to $P(n)$. Then $Q(n-k)$ is true for $n - k \geq 0$ if and only if $P(n)$ is true for $n \geq k$ and we have the modified induction. The practical difference is that we start with k instead of zero.

Example 6.6
Prove that $n^2 \geq 2n$ for all $n \geq 2$.

Proof:

Notice that $2^2 = 4 = 2 \times 2$ so the proposition is true when $n = 2$. We next assume that $P(n)$ is true for some n and we compute:

$$
\begin{aligned}
n^2 &\geq 2n \\
n^2 + 2n + 1 &\geq 2n + 2n + 1 \\
(n+1)^2 &\geq 2n + 2n + 1 \\
(n+1)^2 &\geq 2n + 1 + 1 \\
(n+1)^2 &\geq 2n + 2 \\
(n+1)^2 &\geq 2(n+1)
\end{aligned}
$$

To move from the third step to the fourth step we use the fact that $1 < 2n$ when $n \geq 2$. The last step is $P(n+1)$, which means we have deduced $P(n+1)$ from $P(n)$. Using the modified form of induction we have proved that $n^2 \geq 2n$ for all $n \geq 2$.   □

It is possible to formalize the procedure for using mathematical induction into a three-part process. Once we have a proposition $P(n)$,

(i) First, demonstrate a **base case** by directly demonstrating $P(k)$.

(ii) Next, make the **induction hypothesis** that $P(n)$ is true for some n.

(iii) Finally, we make the **inductive step**, starting with the assumption that $P(n)$ is true, demonstrate $P(n+1)$.

These steps permit us to deduce that $P(n)$ is true for all $n \geq k$.

Example 6.7
Using induction, prove

$$
1 + 2 + \cdots + n = \frac{1}{2}n(n+1)
$$

Proof:

Let $P(n)$ be the statement

$$1 + 2 + \cdots + n = \frac{1}{2}n(n + 1)$$

Base case: $1 = \frac{1}{2}1(1 + 1)$, so $P(1)$ is true.

Induction hypothesis: For some n,

$$1 + 2 + \cdots + n = \frac{1}{2}n(n + 1)$$

Inductive Step: Compute:

$$
\begin{aligned}
1 + 2 + \cdots + (n + 1) &= (1 + 2 + \cdots + n) + (n + 1) \\
&= \frac{1}{2}n(n + 1) + (n + 1) \\
&= \frac{1}{2}(n(n + 1) + 2(n + 1)) \\
&= \frac{1}{2}(n^2 + 3n + 2) \\
&= \frac{1}{2}(n + 1)(n + 2) \\
&= \frac{1}{2}(n + 1)((n + 1) + 1)
\end{aligned}
$$

and so we have shown that if $P(n)$ is true then so is $P(n + 1)$. We have thus proven by mathematical induction that $P(n)$ is true for all $n \geq 1$. $\square$

We now introduce *sigma notation* which makes problems like the one worked in Example 6.7 easier to state and manipulate. The symbol $\sum$ is used to add up lists of numbers. If we wished to sum some formula $f(i)$ over a range from a to b, that is to say $a \leq i \leq b$, then we write:

$$\sum_{i=a}^{b} f(i)$$

On the other hand, if S is a set of numbers and we want to add up $f(s)$ for all $s \in S$ we write:

$$\sum_{s \in S} f(s)$$

The result proved in Example 6.7 may be stated in the following form using sigma notation:

$$\sum_{i=1}^{n} i = \frac{1}{2}n(n+1)$$

Proposition 6.8   Suppose that c is a constant and that $f(i)$ and $g(i)$ are formulas. Then

(i) $\sum_{i=a}^{b} (f(i) + g(i)) = \sum_{i=a}^{b} f(i) + \sum_{i=a}^{b} g(i)$

(ii) $\sum_{i=a}^{b} (f(i) - g(i)) = \sum_{i=a}^{b} f(i) - \sum_{i=a}^{b} g(i)$

(iii) $\sum_{i=a}^{b} c \cdot f(i) = c \cdot \sum_{i=a}^{b} f(i)$.

Proof:

Parts (i) and (ii) are both simply the associative and commutative laws for addition, remembering that subtraction can be thought of as adding the negative of a number. These laws are applied many times allowing us to change the order in which the terms are added and retain the same result. Part (iii) is a similar multiple application of the distributive law $ca + cb = c(a + b)$.   □

The sigma notation lets us work with indefinitely long (and even infinite) sums. There are other similar notations. If $A_1, A_2, \ldots, A_n$ are sets then the intersection or union of all these sets can be written:

$$\bigcap_{i=1}^{n} A_i$$

$$\bigcup_{i=1}^{n} A_i$$

Similarly, if $f(i)$ is a formula on the integers then

$$\prod_{i=1}^{n} f(i)$$

is the notation for computing the product $f(1) \cdot f(2) \cdots\cdot f(n)$. This notation is called **Pi notation**.

When we solve an expression involving $\sum$ to obtain a formula that does not use $\sum$ or "..." as in Example 6.7 then we say we have found a **closed form** for the expression. Example 6.7 finds a closed form for $\sum_{i=1}^{n} i$.

6.1 RECURRENCE RELATIONS

At this point we introduce a famous mathematical sequence in order to create an arena for practicing proofs by induction.

Definition 6.9 The **Fibonacci numbers** are defined as follows. $f_1 = f_2 = 1$ and, for $n \geq 3$, $f_n = f_{n-1} + f_{n-2}$.

Example 6.10 The Fibonacci numbers with four or fewer digits are: $f_1 = 1$, $f_2 = 1$, $f_3 = 2$, $f_4 = 3$, $f_5 = 5$, $f_6 = 8$, $f_7 = 13$, $f_8 = 21$, $f_9 = 34$, $f_{10} = 55$, $f_{11} = 89$, $f_{12} = 144$, $f_{13} = 233$, $f_{14} = 377$, $f_{15} = 610$, $f_{16} = 987$, $f_{17} = 1,597$, $f_{18} = 2,584$, $f_{19} = 4,181$, and $f_{20} = 6,765$.

Example 6.11 Prove that the Fibonacci number f_{3n} is even.

Proof:

Base Case: Notice that $f_3 = 2$, which is even, and so the proposition is true when $n = 1$.

Induction Hypothesis: Assume that the proposition is true for some $n \geq 1$.

Inductive Step: Then

$$
\begin{align}
f_{3(n+1)} &= f_{3n+3} \tag{6.1}\\
&= f_{3n+2} + f_{3n+1} \tag{6.2}\\
&= f_{3n+1} + f_{3n} + f_{3n+1} \tag{6.3}\\
&= 2 \cdot f_{3n+1} + f_{3n} \tag{6.4}
\end{align}
$$

but this suffices because f_{3n} is even by the induction hypothesis while $2 \cdot f_{3n+1}$ is also even. The sum is thus even and so $f_{3(n+1)}$ is even.

It follows by induction that f_{3n} is even for all n. $\square$

For the sake of convenience when we are refering to a sequence of symbols indexed by some integer, such as a_k we will use the notation $\{a_k\}_{k=a}^{b}$ to refer to a sequence of symbols indexed by k where k varies from a to b, with $a \leq b$. The notation $\{a_k\}_{k=a}^{\infty}$ is used to refer to an infinite sequence which starts at $k = a$ and continues for all natural numbers larger than a.

The Fibonacci sequence $\{f_n\}_{n=0}^{\infty}$ is the sequence of the Fibonacci numbers. This sequence is defined in terms of a formula which uses previous terms in the sequence. This is a useful way of defining a sequence and in fact it has a name: A **recurrence relation**.

Definition 6.12   A **recurrence relation** for a sequence $\{a_n\}$ is an equation which expresses a_n in terms of a fixed number of earlier terms of the sequence $a_{n-k}, a_{n-k+1}, \ldots a_{n-1}$, for all integers $n \geq n_0$, where n_0 is a fixed non-negative integer.

The **initial conditions** of a sequence specify the values of the intial terms before the recurrence relation takes effect. In the case of the Fibonacci sequence the initial conditions are $f_1 = f_2 = 1$. A sequence whose terms satisfy a recurrence relation is said to be a *solution* to the recurrence relation. Note that multiple sequences can satisfy the same recurrence relation.

Example 6.13
Consider the sequences $\{a_n\}_{n=0}^{\infty}$ and $\{b_n\}_{n=0}^{\infty}$ given by $a_n = 2a_{n-1} + a_{n-2}$ for $n \geq 2$ with $a_0 = 1, a_1 = 0$, and $b_n = 5 \cdot b_{n-2} + 2 \cdot b_{n-3}$ for $n \geq 3$ with $b_0 = 1, b_1 = 2, b_2 = 2$.

$a_n : 1, 0, 1, 2, 5, 12, 29, \ldots$

$b_n : 1, 2, 2, 12, 14, 64, 94, \ldots$

Both sequences satisfy the recurrence relation $f_n = 5 \cdot f_{n-2} + 2 \cdot f_{n-3}$ for $n \geq 3$. The sequence $\{b_n\}_{n=0}^{\infty}$ does so by definition but $\{a_n\}_{n=0}^{\infty}$ satisfies it as well since

$$a_n = 2a_{n-1} + a_{n-2} = 2(2a_{n-2} + a_{n-3}) + a_{n-2} = 5a_{n-2} + 2a_{n-3}$$

this example demonstrates that multiple sequences may satisfy the same recurrence relation and also a sequence may satisfy multiple recurrence relations.

Mathematical induction is a wonderful tool for proving claims about sequences defined by recurrence relations. Consider the famous *Towers of Hanoi*, a popular puzzle of the late 19th century, invented by the French mathematician E. Lucas. The towers of Hanoi consists of disks of different sizes and a board with three pegs mounted on it. Figure 6.1 displays an example on three disks. The disks are initially placed on the first peg, with the largest disk on the bottom and smaller and smaller disks placed on top. The objective is to move all of the disks on to a different peg, with the largest disk on the bottom and smaller and smaller disks on top but the catch is that you are only permitted to move a single disk at a time from the top of a pile on one peg to another, and may never place a larger disk on top of a smaller disk.

If $\{T_n\}$ is the sequence containing the minimum number of moves needed to solve the puzzle when there are n disks then with some work (see Problem 6.35) it is possible to show that T_n

Towers of Hanoi with three disks

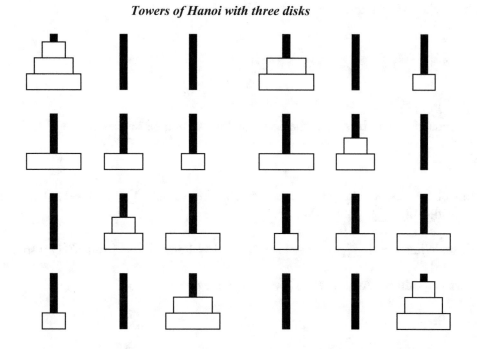

Figure 6.1: Towers of Hanoi.

satisfies the recurrence relation $T_n = 2T_{n-1} + 1$ for $n \geq 2$. From there we can prove the closed form of the solution using mathematical induction.

Example 6.14
Prove that if $T_n = 2T_{n-1} + 1$, for $n \geq 2$ and $T_1 = 1$ then $T_n = 2^n - 1$ for all $n \geq 1$.

Proof by Induction:

Base Case: $T_1 = 1$ and $2^1 - 1 = 1$, thus $T_n = 2^n - 1$ for $n = 1$.

Inductive Hypothesis: Assume that $T_n = 2^n - 1$ for some n.

Inductive Step:

$$T_{n+1} = 2T_n + 1$$
$$T_{n+1} = 2(2^n - 1) + 1$$
$$T_{n+1} = 2^{n+1} - 2 + 1$$
$$T_{n+1} = 2^{n+1} - 1$$

Thus, by mathematical induction the result follows for all $n \geq 1$. $\square$

6.2   ADDITIONAL EXAMPLES

Example 6.15
Prove by induction that the sum of the first n odd whole numbers equals n^2.

Proof:

The sum of the first n odd numbers is given by the formula $\sum_{k=1}^{n}(2k-1)$. Let $P(n)$ be the proposition that $\sum_{k=1}^{n}(2k-1) = n^2$.

Base Case: When $n = 1$, $\sum_{k=1}^{1}(2k-1) = 2(1) - 1 = 1$ and $(1)^2 = 1$ thus $P(1)$ is true.

Inductive Hypothesis: Assume $P(n)$ is true for some n.

Inductive Step:

$$\sum_{k=1}^{n+1}(2k-1) = (2(n+1)-1) + \sum_{k=1}^{n}(2k-1)$$

$$\sum_{k=1}^{n+1}(2k-1) = (2n+2-1) + n^2$$

$$\sum_{k=1}^{n+1}(2k-1) = n^2 + 2n + 1$$

$$\sum_{k=1}^{n+1}(2k-1) = (n+1)^2$$

Thus, $P(n+1)$ is true whenever $P(n)$ is true. By mathematical induction, the result holds for all $n \geq 1$. $\square$

Example 6.16
It is equally or perhaps at times more beneficial to see some examples of incorrect proofs. We provide here some incorrect proofs of a claim and discuss where these proofs fail.

Claim: For $n \geq 1$, $\sum_{k=1}^{n} k(k+1)(k+2) = \frac{n(n+1)(n+2)(n+3)}{4}$.

"Proof 1":

Base Case: $1(2)(3) = 6 = \frac{1(2)(3)(4)}{4}$

Inductive Hypothesis: Assume $P(n)$ is true for some n.

Inductive Step:

$$\sum_{k=1}^{n+1} k(k+1)(k+2) = \frac{(n+1)(n+2)(n+3)(n+4)}{4}$$

So the result holds. $\square$

There are several things wrong with this proof. The Inductive Hypothesis claims $P(n)$ is true for some n, however there was no definition of what the predicate $P(n)$ was supposed to be. While we may make an educated guess in this case the exact nature of the proposition may not be apparent in other circumstances. Note that in Example 6.15 we defined $P(n)$. The biggest problem with the proof however is that in the inductive step we simply state the result we wish to prove. The authors really wish this wasn't a common occurrence on students assignments but it is so common that the authors felt it necessary to point out that stating the result is not a proof. Finally, it's actually necessary to explain why the result holds in general by specifying the fact that we are using mathematical induction.

"Proof 2":

Assume, by way of mathematical induction, that $\sum_{k=1}^{n} k(k+1)(k+2) = \frac{n(n+1)(n+2)(n+3)}{4}$.

Note that:

$$\sum_{k=1}^{n+1} k(k+1)(k+2) = (n+1)(n+2)(n+3) + \frac{n(n+1)(n+2)(n+3)}{4}$$

$$\sum_{k=1}^{n+1} k(k+1)(k+2) = \frac{4(n+1)(n+2)(n+3)}{4} + \frac{n(n+1)(n+2)(n+3)}{4}$$

$$\sum_{k=1}^{n+1} k(k+1)(k+2) = \frac{(n+4)(n+1)(n+2)(n+3)}{4}$$

$$\sum_{k=1}^{n+1} k(k+1)(k+2) = \frac{(n+1)(n+2)(n+3)(n+4)}{4}$$

So by way of mathematical induction, the result follows. $\qquad\square$

This time around the base case of the proof was entirely missing. While this may seem trivial it is in fact not. As we saw earlier multiple distinct sequences may satisfy the same recurrence relation, proving the base case corresponds to proving a proposition on the initial conditions as well. The phrasing "assume, by way of mathematical induction" is a strange way of getting at the inductive hypothesis, and the rest of the proof does not specify that it is dealing with the inductive step (though the actual calculation is perfectly fine). The awkward phrasing is certainly more forgivable than some of the errors in the previous bad proof but it should be restated that a proof is an argument. As such, clarity matters.

"Proof 3":

Let $f(n) = \sum_{k=1}^{n} k(k+1)(k+2)$ and let $g(n) = \frac{n(n+1)(n+2)(n+3)}{4}$.

Base Case: $f(1) = g(1)$

Assume $f(n) = g(n)$ for some $n \geq m$.

$f(m+1) = \sum_{k=1}^{m+1} k(k+1)(k+2)$, so $f(m+1) = (m+1)(m+2)(m+3) + f(m)$ and hence $f(m+1) - f(m) = (m+1)(m+2)(m+3)$.

Now consider $g(m)$ and $g(m+1)$:

$$g(m+1) - g(m) = \frac{(m+1)(m+2)(m+3)(m+4) - m(m+1)(m+2)(m+3)}{4}$$

$$g(m+1) - g(m) = \frac{(m+1)(m+2)(m+3)4}{4} = (m+1)(m+2)(m+3)$$

Since f and g both satisfy the same relation they are equal. Thus, $f(m+1) = g(m+1)$ whenever $f(m) = g(m)$. So by mathematical induction, the result follows. $\qquad\square$

This "Proof" is an interesting case. There's nothing wrong with specifying the separate halves of the equation we want to prove as their own functions f and g. The base case states $f(1) = g(1)$ which is the same as asserting that the result holds when $n = 1$ without showing a calculation to show that it holds. That in itself is an error. The inductive hypothesis is stated but it is not explicitly called an inductive hypothesis. There's nothing actually wrong with that but it makes the argument harder to follow. The fact that the hypothesis is stated "for some $n \geq m$" and that afterward m is used is actually good form that is often overlooked. However, the inductive step is a problem. A bunch of clever manipulation was used to show that essentially $f(m+1) - f(m) = g(m+1) - g(m)$. That is not the same as showing that $g(m+1) = f(m+1)$ in fact it's just demonstrating that f and g satisfy the same recurrence relation. As we've noted before, distinct sequences can satisfy the same recurrence *relation so that does not by itself justify the assumption that f and g are equal. This means that the inductive step did not show that $f(m + 1) = g(m + 1)$ whenever $f(m) = g(m)$, so the proof is invalid.*

6.3   PROBLEMS

Problem 6.17
Suppose that $S = \{a, b, c\}$. Compute and list explicitly the members of the power set, $\mathcal{P}(S)$.

Problem 6.18
Prove that for a finite set X that $|X| \leq |\mathcal{P}(X)|$.

Problem 6.19
Suppose that $X \subseteq Y$ with $|Y| = n$ and $|X| = m$.

Compute the number of subsets of Y that contain X.

Problem 6.20
Compute the following sums:

(i) $\sum_{i=1}^{20} i$,

(ii) $\sum_{i=10}^{30} i$, and

(iii) $\sum_{i=-20}^{21} i$.

Problem 6.21
Compute the sum of the first n positive even numbers.

Problem 6.22
Find a closed form for

$$\sum_{i=1}^{n} i^2 + 3i + 5$$

Problem 6.23
Using mathematical induction, prove the following formulas:

(i) $\sum_{i=1}^{n} 1 = n$,

(ii) $\sum_{i=1}^{n} i^2 = \frac{n(n+1)(2n+1)}{6}$, and

(iii) $\sum_{i=1}^{n} i^3 = \frac{n^2(n+1)^2}{4}$.

Problem 6.24
Let $f(n, 3)$ be the number of subsets of $\{1, 2, \ldots, n\}$ of size 3. Using induction, prove that $f(n, 3) = \frac{1}{6}n(n-1)(n-2)$.

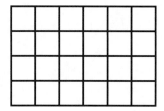

Problem 6.25
Suppose you want to break an $n \times m$ chocolate bar, like the 6×4 example shown above, into pieces corresponding to the small squares shown. What is the minimum number of breaks you can make? Prove your answer is correct.

Problem 6.26
Suppose that we have sets $X_1, X_2, \ldots, X_n$ and $Y_1, Y_2, \ldots, Y_n$ such that $X_i \subseteq Y_i$. Prove that the intersection of all the X_i is a subset of the intersection of all the Y_i:

$$\bigcap_{i=1}^{n} X_i \subseteq \bigcap_{i=1}^{n} Y_i$$

Problem 6.27
Prove by induction that the Fibonacci number f_{4n} is a multiple of 3.

Problem 6.28
Prove by induction that the Fibonacci number f_{5n} is a multiple of 5.

Problem 6.29
Prove that the sequence $\{a_n\}$ is a solution to the recurrence relation $a_n = 2a_{n-1} - a_{n-2}$ if for $n \geq 0$:

(i) $a_n = 1$

(ii) $a_n = 0$

(iii) $a_n = n + 1$

Problem 6.30
Prove that if r is a real number $r \neq 1$ and $r \neq 0$ then

$$\sum_{i=0}^{n} r^i = \frac{1 - r^{n+1}}{1 - r}$$

Problem 6.31
Suppose that $S_1, S_2, \ldots S_n$ are sets. Prove the following generalization of De Morgan's laws:

(i) $\left(\bigcap_{i=1}^{n} S_i\right)^c = \bigcup_{i=1}^{n} S_i^c$ and

(ii) $\left(\bigcup_{i=1}^{n} S_i\right)^c = \bigcap_{i=1}^{n} S_i^c$.

Problem 6.32
Prove by induction that the Fibonacci number f_n has the value

$$f_n = \frac{\sqrt{5}}{5} \cdot \left(\frac{1 + \sqrt{5}}{2}\right)^n - \frac{\sqrt{5}}{5} \cdot \left(\frac{1 - \sqrt{5}}{2}\right)^n$$

Problem 6.33
Prove that $\frac{n(n-1)(n-2)(n-3)}{24}$ is a whole number for any whole number n.

Problem 6.34
Consider the statement "All cars are the same color." and the following "proof."

Proof:

We will prove for $n \geq 1$ that for any set of n cars all the cars in the set have the same color.

- **Base Case:** n=1 If there is only one car then clearly there is only one color the car can be.

- **Inductive Hypothesis:** Assume that for any set of n cars there is only one color.

- **Inductive step:** Look at any set of n + 1 cars. Number them: $1, 2, 3, \ldots, n, n + 1$. Consider the sets $\{1, 2, 3, \ldots, n\}$ and $\{2, 3, 4, \ldots, n + 1\}$. Each is a set of only n cars, therefore for each set there is only one color. But the nth car is in both sets so the color of the cars in the first set must be the same as the color of the cars in the second set. Therefore, there must be only one color among all $n + 1$ cars.

- The proof follows by induction.   □

What is the problem with this proof?

Problem 6.35

Let T_n denote the minimum number of moves needed to solve the Tower of Hanoi. Prove that $T_n = 2T_{n-1} + 1$ for $n \geq 2$. *Without explicitly using the closed form for T_n which we proved using the recurrence relation.*

Hint: Think about how to attack the puzzle from the point of view of when you are allowed to move the largest piece.

CHAPTER 7

Functions

While every reader of this text is likely familiar with functions such as a quadratic function like $f(x) = x^2 + 2x + 1$, this is likely a student's first real introduction to *abstract mathematical functions* (or **set theoretic functions**). In more advanced mathematics the set theoretic definition of functions is used as the default definition of a function.

In this section we will define functions and extend much of our ability to work with sets to infinite sets. There are a number of different types of functions and so this section contains a great deal of terminology.

Recall that two finite sets are the same size if they contain the same number of elements. It is possible to make this idea formal by using functions and, once the notion is formally defined, it can be applied to infinite sets.

7.1   MATHEMATICAL FUNCTIONS

Definition 7.1   An **ordered pair** is a collection of two elements with the added property that one element comes first and one element comes second. The set containing only x and y (for $x \neq y$) is written $\{x, y\}$. The ordered pair containing x and y with x first is written (x, y). Notice that while $\{x, x\}$ is not a well-defined set, (x, x) is a well-defined ordered pair because the two copies of x are different by virtue of coming first and second.

The reason for defining ordered pairs at this point is that it permits us to make an important formal definition that pervades the rest of mathematics.

Definition 7.2   A **function** f with **domain** S and **range** T is a set of ordered pairs (s, t) with first element from S and second element from T that has the property that every element of S appears exactly once in some ordered pair. We write $f : S \to T$ for such a function. We say f is a **mapping** of S to T, or that f maps S to T.

Example 7.3
Suppose that $A = \{a, b, c\}$ and $B = \{0, 1\}$ then

$$f = \{(a, 0), (b, 1), (c, 0)\}$$

is a function from A to B. The function $f : A \to B$ can also be specified by saying $f(a) = 0$, $f(b) = 1$ and $f(c) = 0$.

The set of ordered pairs $\{(a, 0), (b, 1)\}$ is not a function from A to B because c is not the first co-ordinate of any ordered pair. The set of ordered pairs $\{(a, 0), (a, 1), (b, 0), (c, 0)\}$ is not a function from A to B because a appears as the first coordinate of two different ordered pairs.

Set theoretic functions may sometimes be drawn but these drawings look very different from the sort of drawing of a function you may be familiar with from calculus. These drawings are called **function diagrams**. Example 7.4 illustrates the drawing of a set-theoretic function from the set $X = \{1, 2, 3, 4, 5\}$ to the set $Y = \{a, b, c\}$.

Example 7.4

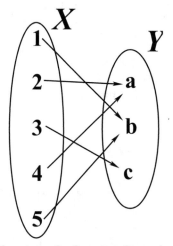

Shown here is an example of a drawing of a function from the set $X = \{1, 2, 3, 4, 5\}$ to the set $Y = \{a, b, c\}$.

The arrows indicate which elements from the two sets are paired together in the ordered pairs of the function.

If we let $f : X \to Y$ be the function shown in the diagram then $f(1) = b$, $f(2) = a$, $f(3) = c$, $f(4) = a$, $f(5) = b$. Alternatively, in the ordered pair notation $f = \{(1, b), (2, a), (3, c), (4, a), (5, b)\}$.

In calculus you may have learned the *vertical line rule* that states that the graph of a function may not intersect a vertical line in more than one point. This corresponds to requiring that each point in the domain of the function appear in only one ordered pair. In set theory, all functions are required to state their domain and range when they are defined. In calculus, functions had a domain and range that were subsets of the real numbers and you were sometimes required to identify the subsets.

We use the symbol $\mathbb{R}$ for the real numbers. We will sometimes use interval notation for contiguous subsets of the reals. For real numbers $a \leq b$

$$
\begin{array}{lll}
(a,b) & \text{is} & \{x : a < x < b\} \\
(a,b] & \text{is} & \{x : a < x \leq b\} \\
[a,b) & \text{is} & \{x : a \leq x < b\} \\
[a,b] & \text{is} & \{x : a \leq x \leq b\}
\end{array}
$$

Example 7.5

This example contrasts the way functions are treated in a typical calculus course with the way we treat them in set theory.

Calculus: Find the domain of the function

$$f(x) = \sqrt{x}$$

Since we know that the square root function exists only for non-negative real numbers the domain is $\{x : x \geq 0\}$.

Set theory: the function $f = \sqrt{x}$ from the non-negative real numbers to the real numbers is the set of ordered pairs $\{(r^2, r) : r \geq 0\}$. This function is well defined because each non-negative real number is the square of some non-negative real number. We can shorten that to $f : [0, \infty) \to \mathbb{R}$ where $f = \{(r^2, r) : r \geq 0\}$.

At times we may find it convenient to write the function in a sort of hybrid notation. We can specify this as the function $f : [0, \infty) \to \mathbb{R}$ given by $f(x) = \sqrt{x}$. This is still in set theory notation as the domain and range are clearly stated, and the specification of the ordered pairs is still clear.

The major contrasts between functions in calculus and functions in set theory are as follows.

1. The domain of functions in calculus are often specified only by implication (you have to know how all the functions used work) and are almost always a subset of the real numbers. The domain in set theory must be explicitly specified and may be any set at all.

2. Functions in calculus typically have graphs that you can draw and look at. Geometric intuition driven by the graphs plays a major role in our understanding of functions. Functions in set theory are seldom graphed and often don't have a graph.

When we have a set theoretic function $f : S \to T$ then the range of f is a subset of T.

Definition 7.6   If f is a function then we denote the domain of f by $dom(f)$ and the range of f by $rng(f)$

Example 7.7

Suppose that $f(n) : \mathbb{N} \to \mathbb{N}$ is defined by $f(n) = 2n$. Then the domain and range of f are the natural numbers: $dom(f) = rng(f) = \mathbb{N}$. If we specify the ordered pairs of f we get

$$f = \{(n, 2n) : n \in \mathbb{N}\}$$

There are actually two definitions of range that are used in mathematics. The definition we are using, the set from which second coordinates of ordered pairs in a function are drawn, is also the definition typically used in computer science. The other definition is the set of second coordinates that actually appear in ordered pairs. This set, which we will define formally later, is the **image** of the function. To make matters even worse the set we are calling the range of a function is also called the **co-domain**. We include these confusing terminological notes for students that may try and look up supplemental material. In general, we prefer to use range to refer to the co-domain, the set the arrow points to in a formal definition of a function (for example the set T when $f : S \to T$ is the function in question).

Definition 7.8   Let X, Y, and Z be sets. The **composition** of two functions $f : X \to Y$ and $g : Y \to Z$ is a function $h : X \to Z$ for which $h(x) = g(f(x))$ for all $x \in X$. We write $g \circ f$ for the composition of g with f.

The definition of the composition of two functions requires a little checking to make sure it makes sense. Since *every* point must appear as a first coordinate of an ordered pair in a function, every result of applying f to an element of X is an element of Y to which g can be applied. This means that h is a well-defined set of ordered pairs. Notice that the order of composition is important—if the sets X, Y, and Z are distinct there is only one order in which composition even makes sense.

Example 7.9

Suppose that $f : \mathbb{N} \to \mathbb{N}$ is given by $f(n) = 2n$ while $g : \mathbb{N} \to \mathbb{N}$ is given by $g(n) = n + 4$. Then

$$(g \circ f)(n) = 2n + 4$$

while

$$(f \circ g)(n) = 2(n + 4) = 2n + 8$$

We now start a series of definitions that divide functions into a number of classes. We will arrive at a point where we can determine if the mapping of a function is reversible, if there is a function that exactly reverses the action of a given function.

Definition 7.10 A function $f : S \to T$ is **injective** or **one-to-one** if no element of T (no second coordinate) appears in more than one ordered pair. Such a function is called an **injection**.

Example 7.11

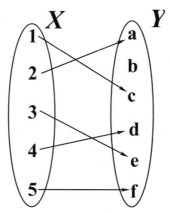

Shown here is an example of a drawing of an injective function from the set $X = \{1, 2, 3, 4, 5\}$ to the set $Y = \{a, b, c, d, e, f\}$.

If we let $f : X \to Y$ be the function shown in the diagram then in the ordered pair notation $f = \{(1, c), (2, a), (3, e), (4, d), (5, f)\}$. As f is an injective function no element in Y appears more than once as the second element in an ordered pair.

Example 7.12
The function $f : \mathbb{N} \to \mathbb{N}$ given by $f(n) = 2n$ is an injection. The ordered pairs of f are $(n, 2n)$ and so any number that appears as a second coordinate does so once.

The function $g : \mathbb{Z} \to \mathbb{Z}$ given by $g(n) = n^2$ is not an injection. To see this notice that g contains the ordered pairs $(1, 1)$ and $(-1, 1)$ so that 1 appears twice as the second coordinate of an ordered pair.

Definition 7.13 A function $f : S \to T$ is **surjective** or **onto** if every element of T appears in an ordered pair. Surjective functions are called **surjections**.

Example 7.14

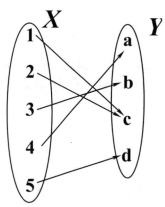

Shown here is an example of a drawing of as surjective function from the set $X = \{1, 2, 3, 4, 5\}$ to the set $Y = \{a, b, c, d\}$.

If we let $f : X \to Y$ be the function shown in the diagram then in the ordered pair notation $f = \{(1, c), (2, c), (3, b), (4, a), (5, d)\}$. As f is a surjective function every element in Y appears at least once as the second element in an ordered pair.

Example 7.15

The function $f : \mathbb{Z} \to \mathbb{Z}$ given by $f(n) = 5 - n$ is a surjection. If we set $m = 5 - n$ then $n = 5 - m$. This means that if we want to find some n so that $f(n)$ is, for example, 8, then $5 - 8 = -3$ and we see that $f(-3) = 8$. This demonstrates that all m have some n so that $f(n) = m$, showing that all m appear as the second coordinate of an ordered pair in f.

The function $g : \mathbb{R} \to \mathbb{R}$ given by $g(x) = \frac{x^2}{1+x^2}$ is not a surjection because $0 \leq g(x) < 1$ for all $x \in \mathbb{R}$.

Definition 7.16   A function that is both surjective and injective is said to be **bijective**. Bijective functions are called **bijections**.

Example 7.17

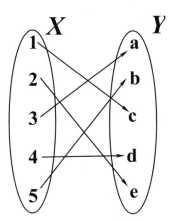

Shown here is an example of a drawing of a bijective function from the set $X = \{1, 2, 3, 4, 5\}$ to the set $Y = \{a, b, c, d, e\}$.

If we let $f : X \to Y$ be the function shown in the diagram then in the ordered pair notation $f = \{(1, c), (2, e), (3, a), (4, d), (5, b)\}$. As f is a bijective function each element in Y appears exactly once as the second element in an ordered pair.

Example 7.18

The function $f : \mathbb{Z} \to \mathbb{Z}$ given by $f(n) = n$ is a bijection. All of its ordered pairs have the same first and second coordinate. This function is called the *identity function*. The function $g : \mathbb{R} \to \mathbb{R}$ given by $g(x) = x^3 - 4x$, shown in Figure 7.1, is not a bijection. It is not too hard to show that it is a surjection, but it fails to be an injection. The portion of the graph shown in Figure 7.1 demonstrates that $g(x)$ takes on the same value more than once. This means that some numbers appear twice as second coordinates of ordered pairs in g. We can figure this out by drawing a graph because g is a function from the real numbers to the real numbers (so a graph makes sense in this case).

For a function $f : S \to T$ to be a bijection every element of S appears in an ordered pair as its first member and every element of T appears in an ordered pair as its second element. Another way to view a bijection is as a matching of the elements of S and T so that every element of S is paired with an element of T. For finite sets this is clearly only possible if the sets are the same size and, in fact, this is the formal definition of "same size" for sets.

Definition 7.19   Two sets S and T are defined to be **the same size** or to have **equal cardinality** if there is a bijection $f : S \to T$.

Example 7.20

The sets $A = \{a, b, c\}$ and $Z = \{1, 2, 3\}$ are the same size. This is obvious because they have the

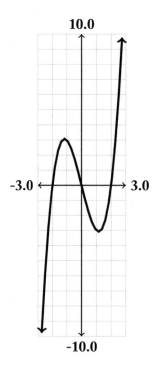

Figure 7.1: The cubic curve $x^3 - 4x$.

same number of elements, $|A| = |Z| = 3$ but we can construct an explicit bijection

$$f = \{(a, 3), (b, 1), (c, 2)\}$$

with each member of A appearing once as a first coordinate and each member of B appearing once as a second coordinate. This bijection is a *witness* that A and B are the same size. In this case it is possible to construct more than one witness. This is often the case.

Let E be the set of even integers. Then the function

$$g : \mathbb{Z} \to E$$

in which $g(n) = 2n$ is a bijection. Notice that each integer can be put into g and that each even integer has exactly one integer that can be doubled to make it. The existence of g is a witness that the set of integers and the set of even integers are the same size. This may seem a bit bizarre because the set $\mathbb{Z} - E$ is the infinite set of odd integers.

In fact, one way to actually define an infinite set is that it is a set which has a proper subset that has the same cardinality. The sets $\mathbb{Z}$ and E are both the same size even though their difference is infinite. We now have an equality test for sizes of infinite sets. We will do a good deal more with the fact that infinite sets have subsets of themselves which are the same size as themselves but not equal, in later chapters.

Bijections have another nice property: they can be unambiguously reversed.

Definition 7.21   The **inverse** of a function $f : S \to T$ is a function $g : T \to S$ so that for all $x \in S$, $g(f(x)) = x$ and for all $y \in T$, $f(g(y)) = y$.

If a function f has an inverse we use the notation f^{-1} for that inverse. Since an exponent of -1 also means reciprocal in some circumstances this can be a bit confusing. The notational confusion is resolved by considering context.

So long as we keep firmly in mind that functions are sets of ordered pairs it is easy to prove the proposition that follows after the next example.

Example 7.22
If E is the set of even integers then the bijection $f(n) = 2n$ from $\mathbb{Z}$ to E has the inverse $g :$ $E \to \mathbb{Z}$ given by $g(2n) = n$. Notice that defining the rule for g as depending on the argument $2n$ seamlessly incorporates the fact that the domain of g is the even integers.

If $f(x) = \frac{x}{x-1}$, shown in Figure 7.2 with its asymptotes $x = 1$ and $y = 1$ then f is a function from the set $H = \mathbb{R} - \{1\}$ to itself. The function was chosen to have asymptotes at equal x and y values; this is a bit unusual. The function f is a bijection. Notice that the graph intersects any horizontal or vertical line in at most one point. Every value except $x = 1$ may be put into f meaning that f is a function on H. Since the vertical asymptote goes off to ∞ in both directions, all values in H come out of f. This demonstrates f is a bijection. This means that it has an inverse which we now compute using a standard technique from calculus classes.

$$
\begin{aligned}
y &= \frac{x}{x-1} \\
y \cdot (x - 1) &= x \\
xy - y &= x \\
xy - x &= y \\
x \cdot (y - 1) &= y \\
x &= \frac{y}{y-1}
\end{aligned}
$$

which tells us that $f^{-1}(x) = \frac{x}{x-1}$ so $f = f^{-1}$: the function is its own inverse.

Proposition 7.23   A function has an inverse if and only if it is a bijection.

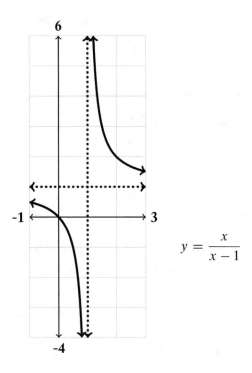

$$y = \frac{x}{x-1}$$

Figure 7.2: The function $f(x) = \frac{x}{x-1}$.

Proof:

Suppose that $f : S \to T$ is a bijection. Then if $g : T \to S$ has ordered pairs that are the exact reverse of those given by f it is obvious that for all $x \in S$, $g(f(x)) = x$, likewise that for all $y \in T$, $f(g(y)) = y$. We have that bijections possess inverses. It remains to show that non-bijections do not have inverses.

If $f : S \to T$ is not a bijection then either it is not a surjection or it is not an injection. If f is not a surjection then there is some $t \in T$ that appears in no ordered pair of f. This means that no matter what $g(t)$ is, $f(g(t)) \neq t$ and we fail to have an inverse. If, on the other hand, $f : S \to T$ is a surjection but fails to be an injection then for some distinct $a, b \in S$ we have that $f(a) = t = f(b)$. For $g : T \to S$ to be an inverse of f we would need $g(t) = a$ and $g(t) = b$, forcing t to appear as the first coordinate of two ordered pairs in g and so rendering g a non-function. We thus have that non-bijections do not have inverses. $\square$

The type of inverse we are discussing above is a **two-sided inverse**. The functions f and f^{-1} are mutually inverses of one another. It is possible to find a function that is a one-way inverse of a function so that $f(g(x)) = x$ but $g(f(x))$ is not even defined. These are called **one-sided inverses**.

Example 7.24
Let $f : [0, \infty) \to \mathbb{R}$ be given by $f(x) = \sqrt{x}$.
Let $g : \mathbb{R} \to [0, \infty)$ be given by $g(x) = x^2$.
Then $g \circ f : [0, \infty) \to [0, \infty)$ is the function $g \circ f(x) = x$.
However, $f \circ g : \mathbb{R} \to \mathbb{R}$ is the function $f \circ g(x) = |x|$.
The function g is a one-sided inverse of f but f is not a one-sided inverse of g.

Proposition 7.25   Suppose that X, Y, and Z are sets. If $f : X \to Y$ and $g : Y \to Z$ are bijections then so is $g \circ f : X \to Z$.

Proof:

This proof is left as an exercise.

Definition 7.26   Suppose that $f : A \to B$ is a function. The **image of A in B** is the subset of B made of elements that appear as the second element of ordered pairs in f. Colloquially, the image of f is the set of elements of B hit by f. We use the notation $Im(f)$ for images. In other words $Im(f) = \{f(a) : a \in A\}$.

The image of a set theoretic function is the range of a function from calculus. In the above definition the set B is the set theoretic range of $f : A \to B$ while $Im(f)$ is the image.

Example 7.27
If $f : \mathbb{N} \to \mathbb{N}$ is given by the rule $f(n) = 3n$ then the set $T = \{0, 3, 6, \ldots\}$ of natural numbers that are multiples of three is the image of f. Notation: $Im(f) = T$.

If $g : \mathbb{R} \to \mathbb{R}$ given by $g(x) = x^2$ then

$$Im(g) = \{y : y \geq 0, y \in \mathbb{R}\}$$

Sometimes it is useful to be able to refer to the set of all ordered pairs drawn from two sets.

Definition 7.28   If A and B are sets then the set of all ordered pairs with the first element from A and the second from B is called the **Cartesian Product** of A and B.

The notation for the Cartesian product of A and B is $A \times B$. Using set notation:

$$A \times B = \{(a,b) : a \in A, b \in B\}$$

Example 7.29
If $A = \{1, 2\}$ and $B = \{x, y\}$ then

$$A \times B = \{(1, x), (1, y), (2, x), (2, y)\}$$

The **Cartesian plane**, the $x - y$ plane that we use to graph functions in calculus, is an example of a Cartesian product of the real numbers with themselves: $\mathbb{R} \times \mathbb{R}$.

7.2   PERMUTATIONS

In this section we will look at a very useful sort of function, bijections of finite sets.

Definition 7.30   A **permutation** is a bijection of a finite set with itself. Likewise a bijection of a finite set X with itself is called a **permutation of X**.

Example 7.31
Let $A = \{a, b, c\}$ then the possible permutations of A consist of the following six functions:

$$\{(a,a)(b,b)(c,c)\} \qquad \{(a,a)(b,c)(c,b)\}$$

$$\{(a,b)(b,a)(c,c)\} \qquad \{(a,b)(b,c)(c,a)\}$$

$$\{(a,c)(b,a)(c,b)\} \qquad \{(a,c)(b,b)(c,a)\}$$

Notice that the number of permutations of three objects does not depend on the identity of those objects. In fact, there are always six permutations of any set of three objects. We now define a handy function that uses a rather odd notation. The method of showing permutations in Example 7.31, explicit listing of ordered pairs, is a bit cumbersome.

Definition 7.32   Assume that we have agreed on an order, e.g., a,b,c, for the members of a set $X = \{a, b, c\}$. Then **one-line** notation for a permutation f consists of listing the first coordinate of the ordered pairs in the agreed on order. The table in Example 7.31 would become:

abc   acb
bac   bca
cab   cba

in one line notation. Notice the saving of space.

There is a formula for counting the number of permutations of n objects.

Definition 7.33   The **factorial** of a natural number n is the product

$$n(n-1)(n-2)\cdots 3\cdot 2\cdot 1 = \prod_{i=1}^{n} i$$

with the convention that the factorial of 0 is 1. We denote the factorial of n as $n!$, spoken "n factorial."

Example 7.34
Here are the first few factorials:

n	0	1	2	3	4	5	6	7
n!	1	1	2	6	24	120	720	5,040

There is an alternate method for defining the factorial of a natural number n, which is worth mentioning. However, before we do there is a notational convention we should mention.

The definition of some functions are naturally broken into cases. For instance, if we wish to define a function $g : \mathbb{Z} \to \mathbb{N}$ as the $g(n) = 2n$ when n is a non-negative integer, but $g(n) = -(2n+1)$ when n is a negative integer, we may use the following convention:

$$g(n) = \begin{cases} 2n & , n \geq 0 \\ -(2n+1) & , n < 0 \end{cases}$$

Definition 7.35   The **factorial** of a natural number n is denoted $n!$, and spoken "n factorial."

$$n! = \begin{cases} 1 & , n = 0 \\ n \cdot (n-1)! & , n > 0 \end{cases}$$

Definition 7.35 is an example of what is known as a *recursive definition*. In a recursive definition an object is defined in terms of smaller instances of the object, while the base cases of the object are explicitly defined by themselves. A recursive definition leverages mathematical induction to specify an object which may be too difficult to specify in a closed form. Recursive definitions only make sense if eventually the nested recursion halts and the definition boils down to base cases that are nested a finite number of times.

Proposition 7.36   The number of permutations of a finite set with n elements is $n!$.

Proof:

This proof is left as an exercise.

Notice that one implication of Proposition 7.25, which states that the composition of bijections is a bijection, is that the composition of two permutations is a permutation. This means that the set of permutations of a set is *closed* under functional composition.

Definition 7.37   A **fixed point** of a function $f : S \to S$ is any $x \in S$ such that $f(x) = x$. We say that **f fixes x**.

7.3   ADDITIONAL EXAMPLES

Example 7.38
Suppose that for finite sets A and B that $f : A \to B$ is a surjective function.

Prove that $|A| \geq |B|$.

Proof:

Let A and B be finite sets, and let $f : A \to B$ be a surjective function.

Since A and B are finite sets, $|A|$ and $|B|$ are simply natural numbers. Since f is surjective we have that for every $b \in B$ there exists $a \in A$ such that $f(a) = b$. Since f is a function each $a \in A$ appears exactly once as the first element of an ordered pair in f. Thus, if for each $b \in B$ we picked an element $a \in A$ such that $f(a) = b$ we would pick exactly $|B|$ elements from A. This means that there is at least one element in the set A for each element in the set B, so $|A| \geq |B|$. □

Example 7.39
Suppose that X and Y are finite sets with $|X| = n$, $|Y| = m$. Count the number of functions from X to Y.

Proof:

Let X and Y be finite sets with $|X| = n$, $|Y| = m$. A function $f : X \to Y$ is a set of ordered pairs which is a subset of $X \times Y$ that has the property that each element in X appears as the first element in exactly one ordered pair.

For each element $a \in X$ there are m choices for $f(a)$ corresponding to the m elements of Y. There are exactly n elements in X. This means there are m choices to be made n times. Thus, there are m^n different functions from X to Y. $\square$

Note: The preceding examples are proofs where we utilize counting techniques. They are provided as an example of how to do such proofs. These sorts of proofs tend to show up when it's necessary to deal with results which explicitly specify or compare natural numbers. Students are encouraged to reference Chapter 1 if they need a reminder on the basic counting techniques before tackling the problems.

Example 7.40
Prove that Definition 7.33 and Definition 7.35 of factorial are equivalent.

Proof:

We will show that Definitions 7.33 and 7.35 are equivalent by using mathematical induction.

Let $f(n) = \prod_{i=1}^{n} i$ and let $g(n) = n \cdot g(n-1)$ with the convention that $f(0) = g(0) = 1$.

Base Case: $f(0) = 1 = g(0)$, by definition for both f and g.

Inductive Hypothesis: Suppose that $f(m) = g(m)$ for some $m \in \mathbb{N}$.

Inductive Step:

$$f(m+1) = \prod_{i=1}^{m+1} i$$

$$f(m+1) = (m+1) \cdot \prod_{i=1}^{m} i$$

$$f(m+1) = (m+1) \cdot f(m)$$

$$f(m+1) = (m+1) \cdot g(m)$$

By definition $g(m+1) = (m+1) \cdot g(m)$ for any $m \in \mathbb{N}$, so $f(m+1) = g(m+1)$.

Thus, by mathematical induction it follows that $f(n) = g(n)$ for all $n \in \mathbb{N}$. $\square$

Example 7.41
If f is a permutation of a finite set, prove that the sequence $f, f \circ f, f \circ (f \circ f), \ldots$ must contain repeated elements.

Interlude

The Collatz Conjecture One of the most interesting features of mathematics is that it is possible to phrase problems in a few lines that turn out to be incredibly hard. The Collatz conjecture was first posed in 1937 by Lothar Collatz. Define the function f from the natural numbers to the natural numbers with the rule

$$f(n) = \begin{cases} 3n + 1 & \text{, when n is odd} \\ \frac{n}{2} & \text{, when n is even} \end{cases}$$

The Collatz conjecture is that if you apply f repeatedly to a positive integer then the resulting sequence of numbers eventually arrives at one. If we start with 17, for example, the result of repeatedly applying f is:

$$f(17) = 52, f(52) = 26, f(26) = 13, f(13) = 40, f(40) = 20, f(20) = 10,$$
$$f(10) = 5, f(5) = 16, f(16) = 8, f(8) = 4, f(4) = 2, f(2) = 1$$

The sequences of numbers generated by repeatedly applying f to a natural number are called *hailstone sequences* with the collapse of the value when a large power of 2 appears being analogous to the impact of a hailstone. If we start with the number 27 then 111 steps are required to reach one and the largest intermediate number is 9232. This quite irregular behavior of the sequence is not at all apparent in the original phrasing of the problem.

The Collatz conjecture has been checked for numbers up to 5×2^{61} (about 5.764×10^{18}) by using a variety of computational tricks. It has not, however, been proven or disproven. The very simple statement of the problem causes mathematicians to underestimate its difficulty. At one point a mathematician suggested that the problem might have been developed by the Russians as a way to slow American mathematical research. This was after several of his colleagues spent months working on the problem without obtaining results.

A simple (but incorrect) argument suggests that hailstone sequences ought to grow indefinitely. Half of all numbers are odd, half are even. The function f slightly more than triples odd numbers and divides even numbers in half. Thus, on average, f increases the value of numbers. The problem is this: half of all even numbers are multiples of four and so are divided in half twice. One-quarter of all even numbers are multiples of eight and so get divided in half three times, and so on. The net effect of factors that are powers of two is to defeat the simple argument that f grows "on average."

Proof:

Let S be a finite set with n elements. Let $f : S \to S$ be a permuation of S.

Proposition 7.25 tells us that the composition of f and any arbitrary bijection $h : S \to S$ is a bijection. Since f and h are bijections mapping S to S they are permutations, and hence their composition is a permutation. Note that there are only a finite number of permutations of an n element set (there are in fact exactly $n!$ permutations). This means that when repeatedly composing f with itself, the sequence $f, f \circ f, f \circ (f \circ f), \dots$ must eventually contain repeated elements, because each element in the sequence is a permutation and the sequence is infinite. □

Example 7.42
Prove the composition of injections is an injection.

Proof:

Let X, Y, and Z be sets, and suppose $f : X \to Y$, and $g : Y \to Z$ are injective functions. Let $h : X \to Z$ be given by $h(x) = g \circ f(x)$, $\forall x \in X$.

We need to show that no element in Z appears in more than one ordered pair of h. Suppose that for some $a, b \in X$, we have that $h(a) = h(b)$. This means that $g(f(a)) = g(f(b))$, since g is an injective function this implies that $f(a) = f(b)$, since no element of Z appears in more than one ordered pair in g. Since f itself is an injective function we have that $f(a) = f(b)$ implies that $a = b$. Thus, we have that if $h(a) = h(b)$ then $a = b$. This means that no element of Z appears in more than one ordered pair of h, thus h is an injective function. This demonstrates that the composition of injections is an injection. □

7.4   PROBLEMS

Problem 7.43
Suppose for finite sets A and B that $f : A \to B$ is an injective function. Prove that

$$|B| \geq |A|$$

Problem 7.44
Give the definition of a surjective function using the image of a function.

Problem 7.45
Using functions from the integers to the integers give an example of

(i) A function that is an injection but not a surjection.

(ii) A function that is a surjection but not an injection.

(iii) A function that is neither an injection nor a surjection.

(iv) A bijection that is not the identity function.

Problem 7.46
True or false (and explain): The function $f(x) = \frac{x-1}{x+1}$ is a bijection from the real numbers to the real numbers.

Problem 7.47
Find a function that is an injection of the integers into the even integers that does not appear in any of the examples in this chapter.

Problem 7.48
For each of the following functions from the real numbers to the real numbers say if the function is surjective or injective. It may be neither.

$$\text{(i) } f(x) = x^2 \quad \text{(ii) } g(x) = x^3 \quad \text{(iii) } h(x) = \begin{cases} \sqrt{x} & , x \geq 0 \\ -\sqrt{-x} & , x < 0 \end{cases}$$

Problem 7.49
Suppose that $B \subset A$ and that there exists a bijection $f : A \to B$. What may be reasonably deduced about the sets A and B?

Problem 7.50
Suppose that A and B are finite sets. Prove that $|A \times B| = |A| \cdot |B|$.

Problem 7.51
Suppose that we define $h : \mathbb{N} \to \mathbb{N}$ as follows. If n is even then $h(n) = n/2$ but if n is odd then $h(n) = 3n + 1$. Determine if h is a (i) surjection or (ii) injection.

Problem 7.52
Suppose the Cartesian product $A \times B$ is a function. What can you deduce about the sets A and B? Justify your answer.

Hint: There are two cases.

Problem 7.53
Prove Proposition 7.25.

Problem 7.54
Prove or disprove: the composition of surjections is a surjection.

Problem 7.55
Prove Proposition 7.36.

Problem 7.56
List all permutations of $X = \{1, 2, 3, 4\}$ using one-line notation.

Problem 7.57
Suppose that X is a set and that f, g, and h are permutations of X. Prove that the equation $f \circ g = h$ has a solution g for any given permutations f and h.

Problem 7.58
Examine the permutation f of $Q = \{a, b, c, d, e\}$ which is **bcaed** in one line notation. If we create the series f, $f \circ f$, $f \circ (f \circ f)$, ... does the identity function, **abcde**, ever appear in the series? If so, what is its first appearance? If not, why not?

Problem 7.59
Suppose that X and Y are finite sets and that $|X| = |Y| = n$. Prove that there are $n!$ bijections of X with Y.

Problem 7.60
Suppose that X and Y are sets with $|X| = n$, $|Y| = m$ for $m > n$. Count the number of injections of X into Y.

Problem 7.61
For a finite set S with a subset T prove that the permutations of S that have all members of T as fixed points form a set that is closed under functional composition.

Problem 7.62
Compute the number of permutations of a set S with n members that fix at least $m < n$ points.

Problem 7.63

Using any technique at all, estimate the fraction of permutations of an n-element set that have no fixed points. This problem is intended as an exploration.

Problem 7.64

Let X be a finite set with $|X| = n$. Let $C = X \times X$. How many subsets of C have the property that every element of X appears once as a first coordinate of some ordered pair and once as a second coordinate of some ordered pair?

Problem 7.65

An alternate version of Sigma ($\sum$) and Pi ($\prod$) notation works by using a set as an index. So if $S = \{1, 3, 5, 7\}$ then

$$\sum_{s \in S} s = 16 \text{ and } \prod_{s \in S} s = 105$$

Given all the material so far, give and defend reasonable values for the sum and product of an empty set.

Problem 7.66

Suppose that $f_\alpha : [0, 1] \to [0, 1]$ for $-1 < \alpha < \infty$ is given by

$$f_\alpha(x) = \frac{(\alpha + 1)x}{\alpha x + 1},$$

prove that f_α is a bijection.

CHAPTER 8

The Integers and Beyond

God made the integers, all else is the work of man.
- *Leopold Kronecker [1886]*

8.1 DIVISION ON THE INTEGERS

We begin with a technical property that is useful in proofs. While the principle applies to the integers it is given in a more general context that includes the real numbers.

Proposition 8.1   The Trichotomy Principle
A real number r is positive, negative, or equal to zero.

We now move on to divisibility, one of the most interesting properties of the integers.

Definition 8.2   We say that an integer m **divides** an integer n if there is an integer q so that

$$n = qm$$

If m divides n we use the notation $m|n$. If m fails to divide m we say $m \nmid n$. If $m|n$ we also say that n *is a multiple of m*.

Definition 8.3   Suppose that n is an integer. We say n is **even** if $2|n$, we say n is **odd** if $2 \nmid n$. We say n is **doubly even** if $4|n$ and **triply even** if $8|n$.

The notion of divisibility leads fairly directly to the following useful proposition.

Theorem 8.4   Division Theorem (Euclid's Division Theorem)
*If a and b are positive integers then there are unique integers q and $0 \le r < b$ for which $a = bq + r$. We call q the **quotient** of a divided by b and r the **remainder** of a divided by b.*

Proof:

Consider the set of all non-negative integers

$$S = \{a - bq : (a - bq) \ge 0 \text{ and } q \in \mathbb{Z}\}$$

Since q may be positive or negative, this set is not empty. By the Well-Ordering Principle it has a least member which we name r. We now choose q to be the specific value so that $a - bq = r$ or, after algebraic rearrangement $a = bq + r$. From the definition of S we see that $r \geq 0$. If $r > b$ then we see that $a = b(q + 1) + (r - b)$ with $r - b > 0$. Which means that $r - b \in S$, which is impossible because r is defined to be the smallest element of S. We thus have the condition that $0 \leq r < b$. It remains to demonstrate uniqueness of q and r.

Suppose that

$$a = bq_1 + r_1 = bq_2 + r_2$$

then

$$
\begin{aligned}
bq_1 + r_1 &= bq_2 + r_2 \\
b(q_1 - q_2) &= (r_2 - r_1)
\end{aligned}
$$

This implies that $b|(r_2 - r_1)$. Since r_1 and r_2 are both non-negative and smaller than b it follows that $-b < r_2 - r_1 < b$. The only such number between $-b$ and b that is divisible by b is zero, so $r_2 - r_1 = 0$ and we see $r_1 = r_2$. From this we may deduce that $b(q_1 - q_2) = 0$. Since $b > 0$ this in turn implies that $q_1 - q_2 = 0$ and so $q_1 = q_2$. Thus, the values of q and r are unique. $\square$

The above proof uses a property of the integers that we will now state explicitly and explore later. If $ab = 0$ then $a = 0$ or $b = 0$. This property also holds for other number systems you may be familiar with, such as the rational numbers, the real numbers, and the complex numbers. Later we will give examples of number systems that do not have this property.

Example 8.5
Apply the division theorem to the numbers $a = 91$ and $b = 11$.

Solution:

$91 = 11 \cdot 8 + 3$. This means that the quotient of 91 divided by 11 is 8 while the remainder of 91 divided by 11 is 3.

Proposition 8.6   If $a|b$ and $b|c$ then $a|c$. Another way of saying this is to state that *divisibility is transitive*.

Proof:

This proof is left as an exercise.

Definition 8.7   A **common divisor** of two integers a and b is an integer c so that $c|a$ and $c|b$.

Definition 8.8   The **greatest common divisor** of two non-zero integers a and b is a common divisor g with the added properties that g is positive and that for any common divisor c of a and b, we have that $c|g$. The notation for the greatest common divisor of a and b is $GCD(a,b)$.

Example 8.9

The positive common divisors of the number 96 and 54 are 1, 2, 3, and 6. The greatest common divisor is thus 6. The positive common divisors of 105 and 98 are 1 and 7. The greatest common divisor is 7.

Notice that while we define the greatest common divisor as being the positive common divisor that is divisible by all other common divisors it *happens* to also be the numerically largest.

Theorem 8.10

Any two non-zero integers a and b have a unique positive greatest common divisor.

Proof:

Without loss of generality, assume a and b are positive (the resulting arguments hold if a is replaced with $-a$ when a is negative, and b is replaced with $-b$ when b is negative). Let $I = \{ax + by \geq 0 : x, y \in \mathbb{Z}\}$. Since x, y may take on any integer value it is clear that I is non-empty and since $a, b \neq 0$, I must contain a least positive element $d = au + bv$ for some integers u and v. This follows from the Well-Ordering Principle. Using the division theorem, set $a = qd + r = q(au + bv) + r$. Then we see that $a(1 - qu) + b(-qv) = r$, implying that $r \in I$. This in turn tells us that $r \geq d$ or $r = 0$ since d was defined to be the least positive element in I. Note that $r \geq d$ is impossible because the division theorem forces $0 \leq r < d$, therefore $r = 0$. This yields, at last, the result that $a = qd$ and hence $d|a$. By the same argument with b in place of a we see that $d|b$. We deduce that d is a common divisor of a and b.

Suppose that c is any common divisor of a and b. Then $a = cn$ and $b = cm$ for some integers n and m. Compute for arbitrary integers x and y:

$$ax + by = cnx + cmy$$
$$= c(nx + my)$$

We deduce that c divides any member of I and hence divides d. It follows that all common divisions of a and b divide d and so d is the greatest common divisor of a and b. Suppose a and b have two greatest positive common divisors. Then each divides the other, forcing them to be equal. To see this notice that if $t|s$ for positive t and s it must be the case that $t \leq s$. We thus have uniqueness.   □

Definition 8.11   If a, b, x, and y are integers then we call $ax + by$ a **linear combination of** a **and** b with **coefficients** x and y.

The following is drawn directly from the proof of Theorem 8.10.

Corollary 8.12   The smallest positive linear combination of a and b, where $a, b \neq 0$, is $GCD(a, b)$.

Corollary 8.13   A common divisor of a and b divides every linear combination of a and b.

The proof of this corollary is embedded in the proof of Theorem 8.10. There is a famous algorithm that permits rapid computation of the greatest common divisor of two positive integers.

Algorithm 8.14   Euclid's Algorithm
Assume $a > b > 0$. Set $a = r_{-1}$ and $b = r_0$. The following series of equations:

$$
\begin{aligned}
a &= bq_1 + r_1 \\
b &= r_1 q_2 + r_2 \\
r_1 &= r_2 q_3 + r_3 \\
&\cdots \\
r_i &= r_{i+1} q_{i+1} + r_{i+2} \\
&\cdots \\
r_{n-2} &= r_{n-1} q_n + 0
\end{aligned}
$$

eventually produces a remainder $r_n = 0$. In this case $r_{n-1} = GCD(a, b)$.

Proof:

Notice that the remainders have the property that $r_i > r_{i+1} \geq 0$. If a remainder is positive then the next remainder can be computed, thus the process stops when a zero remainder is found. Since a and b are finite, it follows that a finite number of remainders are computed. We see from this that there is a smallest positive remainder, r_{n-1}. Since $a = r_{-1}$ and $b = r_0$ each remainder from the first one (r_1) is a linear combination of the two remainders preceding it and hence all are linear combinations of a and b. This means, by Corollary 8.13, that every divisor of a and b divides each of the remainders. So the greatest common divisor divides each of the remainders. Therefore, $GCD(a, d) \leq r_k$ for all $k < n$.

Notice that the last equation shows that r_{n-1} divides r_{n-2}. Since r_{n-3} is a linear combination of r_{n-1} and r_{n-2}, we see r_{n-1} also divides r_{n-3}. Following the equations back in like fashion we see that r_{n-1} divides all the remainders and, as we reach the top of the stack of equations, a and b as well. This means that r_{n-1} is a common divisor of a and b. However, this also means that $r_{n-1} \leq GCD(a, b)$.

Therefore, $GCD(a, b) \leq r_{n-1} \leq GCD(a, b)$ and thus r_{n-1} is in fact the greatest common divisor of a and b, and the algorithm works as stated.   □

Notice that since $r_{n-1} = GCD(a, b)$ is it possible, by substituting up the chain of equations created by applying Euclid's algorithm, to actually find coefficients x and y so that $GCD(a, b) = ax + by$.

Example 8.15
Using Euclid's algorithm, find x and y so that

$$GCD(5, 8) = x \cdot 5 + y \cdot 8$$

Solution:

$$
\begin{aligned}
8 &= 1 \cdot 5 + 3 \\
5 &= 1 \cdot 3 + 2 \\
3 &= 1 \cdot 2 + 1 \\
2 &= 1 \cdot 1 + 1 \\
1 &= 1 \cdot 1 + 0
\end{aligned}
$$

And so we see that $GCD(5, 8) = 1$. We now run the calculations backward, substituting in for the smallest term in each step.

$$
\begin{aligned}
1 &= 2 - 1 \\
&= 2 - (3 - 2) \\
&= 2 \cdot 2 - 3 \\
&= 2 \cdot (5 - 3) - 3 \\
&= 2 \cdot 5 - 3 \cdot 3 \\
&= 2 \cdot 5 - 3 \cdot (8 - 5) \\
&= 5 \cdot 5 - 3 \cdot 8
\end{aligned}
$$

Verify: $GCD(5, 8) = 1 = 25 - 24$. So we see $x = 5$ and $y = -3$ are the required coefficients.

8.2   PRIME NUMBERS AND PRIME FACTORIZATION

Definition 8.16   A **prime number** is a natural number larger than 1 whose only positive divisors are itself and 1.

Example 8.17
The primes with three or fewer digits are:

2, 3, 5, 7, 11, 13, 17, 19, 23, 29, 31, 37, 41, 43, 47, 53, 59, 61, 67, 71, 73, 79, 83, 89, 97, 101, 103, 107, 109, 113, 127, 131, 137, 139, 149, 151, 157, 163, 167, 173, 179, 181, 191, 193, 197, 199, 211, 223, 227, 229, 233, 239, 241, 251, 257, 263, 269, 271, 277, 281, 283, 293, 307, 311, 313, 317, 331, 337, 347, 349, 353, 359, 367, 373, 379, 383, 389, 397, 401, 409, 419, 421, 431, 433, 439, 443, 449, 457, 461, 463, 467, 479, 487, 491, 499, 503, 509, 521, 523, 541, 547, 557, 563, 569, 571, 577, 587, 593, 599, 601, 607, 613, 617, 619, 631, 641, 643, 647, 653, 659, 661, 673, 677, 683, 691, 701, 709, 719, 727, 733, 739, 743, 751, 757, 761, 769, 773, 787, 797, 809, 811, 821, 823, 827, 829, 839, 853, 857, 859, 863, 877, 881, 883, 887, 907, 911, 919, 929, 937, 941, 947, 953, 967, 971, 977, 983, 991, and 997

Proposition 8.18   If p is prime and m, n are positive integers such that $p|mn$ then either $p|n$ or $p|m$.

Proof:

If $p|m$ then we we are done so assume that $p \nmid m$. In this case $GCD(p, m) = 1$ and Corollary 8.12 tells us that we can find x and y so that $1 = mx + py$. Multiplying this expression through by n yields $n = nmx + npy$. Since $p|nm$ we have that for some c, $pc = nm$ and so we may substitute to obtain $n = pcx + npy = p(cx + ny)$ which implies that $p|n$. $\square$

This proposition is used to prove a famous theorem.

Theorem 8.19   (Unique Factorization Theorem)
A positive integer n greater than 1 may be factored into prime numbers, some of which may be repeated, uniquely up to the order of the factors.

Proof:

Let S be the set of positive integers greater than 1 which fail to factor into primes. By well ordering, if it is non-empty then it has a least member m. If m is a prime then it factors into itself, yielding a factorization into primes. If m is not a prime it follows that m has two factors, a and b, larger than 1 so that $m = ab$. The minimality property of m ensures that a and b both factor into primes. But the product of the prime factorizations of a and b is equal to m and so forms a prime factorization of m. It follows that our hypothesis that S is non-empty is false and all positive integers greater than 1 factor into prime numbers. It remains to demonstrate uniqueness.

Suppose that $n = p_1 p_2 \cdots p_k = q_1 q_2 \cdots q_l$ are two prime factorizations of n. Then by Proposition 8.18 we see that $p_1 | q_1 q_2 \cdots q_l$ which implies that $p_1 = q_i$ for some i. We renumber the members of the second prime factorization so that $p_1 = q_1$. This tells us that $p_2 \cdots p_k = q_2 \cdots q_l$

and we may continue in like fashion obtaining both $p_i = q_i$ and, as we eliminate primes in pairs, that $k = l$. It follows that the factorization into primes is unique up to the order of the prime factors. $\square$

Example 8.20

Here are some examples of prime factorizations:

$$24 = 2^3 \cdot 3$$
$$54 = 2 \cdot 3^3$$
$$91 = 7 \cdot 13$$
$$105 = 3 \cdot 5 \cdot 7$$
$$256 = 2^8$$
$$625 = 5^4$$
$$1,000 = 2^3 \cdot 5^3$$
$$5,040 = 2^4 \cdot 3^2 \cdot 5 \cdot 7$$
$$30,030 = 2 \cdot 3 \cdot 5 \cdot 7 \cdot 11 \cdot 13$$

Theorem 8.21   Euclid's Theorem

There are an infinite number of prime numbers.

Proof:

Assume, by way of contradiction, that there are a finite number of prime numbers. Note that 2 is a prime number, since it is a natural number larger than 1 whose only positive divisors are itself and 1.

Let P be the set of prime numbers. Since we know $2 \in P$, P is non-empty. Let $n = |P|$ and denote the n prime numbers as p_k for $1 \le k \le n$.

Consider the number $q = (\prod_{k=1}^{n} p_k) + 1$ and whether or not q is prime.

If q is prime, then note that $q > \prod_{k=1}^{n} p_k$ and hence $q > p_k$ for all k, so $q \notin P$ and hence P is does not contain all prime numbers, which is a contradiction. If q is not prime then, by the unique factorization theorem, it has a prime divisor, a prime number m such that $m|q$. However, all prime numbers are in the product $(\prod_{k=1}^{n} p_k)$, which means that $m|(\prod_{k=1}^{n} p_k)$. This means that $m|q - (\prod_{k=1}^{n} p_k)$. However, $q - (\prod_{k=1}^{n} p_k) = 1$. Thus, $m|1$ and $m \le 1$ while

m is a prime number, which is a contradiction. Thus, whether q is a prime number or not it leads to a contradiction. Thus, there cannot be a finite number of prime numbers.   □

Definition 8.22   If two numbers have no prime factors in common we call those numbers **relatively prime**.

Proposition 8.23   Two numbers a and b are relatively prime iff $GCD(a, b) = 1$.

Proof:

($\Rightarrow$) Suppose that a and b are relatively prime. Any common divisor of a and b must divide $GCD(a, b)$. Since there are no common prime divisors, this tells us that $GCD(a, b)$ has no prime divisors. This forces $GCD(a, b) = 1$ since this is the only positive integer with an empty set of prime divisors.

($\Leftarrow$) Suppose, on the other hand, that $GCD(a, b) = 1$. Any common divisor of a and b must divide the GCD and so we see there are no common divisors larger than 1. This eliminates the possibility of common prime divisors, and we see that a and b are relatively prime.   □

Definition 8.24   A **common multiple** of integers a and b is a number c such that $a|c$ and $b|c$. The **least common multiple** of a and b is a positive common multiple of a and b that divides all other common multiples. The notation for the least common multiple of a and b is $LCM(a, b)$.

Example 8.25
Consider $LCM(5, 7) = 35$, $LCM(24, 18) = 72$, and $LCM(25, 625) = 625$, in terms of the prime factorization of the numbers involved.

Proposition 8.26   For any two natural numbers a and b,

$$ab = GCD(a, b) \cdot LCM(a, b)$$

Proof:

This proof is left as an exercise.

Interlude
A Special Type of Prime Number

Large prime numbers are valuable. One of the most common types of secure encryption, the RSA algorithm, relies on pairs of large prime numbers. There is also no known algorithm for generating large prime numbers that does not require extensive testing. There is a family of numbers which are rich in primes, at least for small values of the family.

Definition 8.27   A **Mersenne prime** is a prime number of the form $2^p - 1$ where p is itself prime.

For $2^p - 1$ to be prime, p itself must be prime, a fact you are asked to prove in the exercises. Unfortunately, not all number of the form $2^p - 1$ are prime. The first few examples are:

prime	$2^p - 1$	factorization
2	3	3
3	7	7
5	31	31
7	127	127
11	2,047	23×89
13	8,191	8,191
17	131,071	131,071
19	524,287	524,287
23	8,388,607	$47 \times 178,481$

As of May 2020 a total of 51 Mersenne primes have been discovered, the largest of which has 24,862,047 digits—this number is also the largest prime number currently known. There may be Mersenne primes smaller than the current largest known. Another exercise in these notes asks you to prove that for a prime $q > 2$ the number $q^n - 1$ is never prime (for $n > 1$).

Prime numbers are still located by checking to see if they factor, but directly testing all primes that might be factors of a number large enough to be a new, useful prime is too time consuming for most practical applications (like data encryption). Tests for primality called *probabilistic tests* are used which do not demonstrate that a number is prime. Instead they estimate a probability that the number is not prime. If this probability is small enough, the number is used as a prime.

8.3   ADDITIONAL EXAMPLES

Example 8.28   Use Euclid's algorithm to determine the $GCD(325, 45)$.

Solution:

$$325 = 7(45) + 10$$
$$45 = 4(10) + 5$$
$$10 = 2(5) + 0$$

So $GCD(325,45) = 5$.

Example 8.29
For positive integers a and b, let $A = \{an : n \in \mathbb{N}\}$ and let $B = \{bn : n \in \mathbb{N}\}$. Prove that $LCM(a,b)$ is the smallest non-zero member of $A \cap B$.

Proof:

Let a and b be positive integers and let $A = \{an : n \in \mathbb{N}\}$ and $B = \{bn : n \in \mathbb{N}\}$. Since A and B are sets of non-negative multiples of a and b, respectively, it follows that $A \cap B$ is the set of all non-negative common multiples of a and b.

Consider the set $A \cap B - \{0\}$, it is non-empty since A and B contain all non-negative multiples of a and b, respectively, and a and b are positive. Thus, by the Well-Ordering Principle, $A \cap B - \{0\}$ contains a smallest positive element d. Let $m, k \in \mathbb{N}$ such that $d = am$ and $d = bk$.

Let z be any positive common multiple of a and b, with $z = ax$ and $z = by$ for some $x, y \in \mathbb{N}$. By the unique factorization theorem d can be factored into primes. Each of these prime numbers p divides d, and so must either divide a or divide m, and also must divide b or divide k. If it divides a or b then it divides z. Assume by way of contradiction that p does not divide either a or b then the number v given by multiplying together all prime factors of a and b (including repeated prime factors), is a smaller number than d, since it does not include p, which is a common multiple of both a and b. This is a contradiction. Thus, each prime factor of d divides either a or b and hence must divide every common multiple of a and b. Moreover, each time a prime factor is repeated in d it must be repeated at least that many times in a common multiple of a and b or else d would again be larger than the common multiple. This means that d is made up of the prime factors of a and b repeated the absolute minimum number of times necessary to divide any common multiple of a and b. Thus, the numerically smallest non-zero member of $A \cap B$ is in fact the $LCM(a,b)$. □

8.4   PROBLEMS

Problem 8.30
Apply the division theorem to find the quotient and remainder for each of the following pairs of numbers a, b:

(i) 14, 5   (ii) 110, 36   (iii) 75, 15   (iv) 4, 15   (v) 111, 38

Problem 8.31
Find GCD(a,b) with Euclid's algorithm for each of the following pairs of numbers:

(i) 70, 63   (ii) 91, 13   (iii) 41, 32   (iv) 89, 55   (v) 233, 144

Problem 8.32
For each pair (a, b) in Problem 8.31 find values x and y so that $GCD(a, b) = ax + by$.

Problem 8.33
Looking at the computations you did in Problem 8.31 can you find a collection of pairs of numbers that take a large number of steps for Euclid's algorithm to resolve, relative to the size of the numbers?

Problem 8.34
Are the two sets "the positive multiples of 3" and "the positive multiples of 5" the same size?

Problem 8.35
Prove Proposition 8.6.

Problem 8.36
Prove, for integers a and b, that if $a|b$ and $b|a$ then $a = \pm b$.

Problem 8.37
Find the prime factorization of the following numbers:

(i) 91   (ii) 255   (iii) 656   (iv) 39,269   (v) 65,535

Problem 8.38
Prove Proposition 8.26.

Problem 8.39
Suppose that $2^n - 1$ is prime. Prove that n is prime.

Problem 8.40
Suppose that n is a natural number smaller than 289. Prove that n is prime iff $GCD(30030, n) = 1$.

Problem 8.41
It is possible to generalize the notion of linear combination to any number of terms. For example, if all the symbols represent integers, then the formula $ax + by + cz$ is a linear combination of a, b, and c.

Define $GCD(a, b, c)$ to be a common divisor of a, b, and c that is divisible by all other common divisors of a, b, and c.

Prove that $GCD(a, b, c)$ is the smallest positive linear combination of a, b, and c.

Problem 8.42
Suppose that $GCD(a, b) = 75$ and $LCM(a, b) = 2{,}250$, find the value of the product ab.

Problem 8.43
Suppose that $GCD(a, b) = 75$ and $LCM(a, b) = 2{,}250$. Can the value of a and b be uniquely determined? Prove your answer.

Problem 8.44
Let $f(n) = n(n - 1)(n - 2)$. Prove that $f(n)$ is a multiple of six for all n.

Problem 8.45
Let $f(n) = n(n - 1)(n - 2)(n - 3)$. Prove that $f(n)$ is triply even for all integers n.

Problem 8.46
Suppose that $f(n) = n(n + 1)(n + 2)(n + 3)(n + 4)$. Find the largest whole number that is a divisor of $f(n)$ for all n.

Problem 8.47

Suppose that

$$g(n) = \prod_{i=0}^{m}(n+i)$$

Find the smallest value of m that ensures that $256|g(n)$ for all n.

Problem 8.48

The **Sieve of Eratosthenes** is an algorithm for finding prime numbers that works as follows.

List the numbers from 2 to n. Repeat the following steps. Circle the smallest number that is neither circled nor crossed out. Cross out all multiples of the number you just circled. Repeat these steps until all numbers are circled or crossed out. The circled numbers are the primes less than or equal to n.

Apply the sieve for $n = 100$.

Problem 8.49

For $2 \le n \le 12$ find the prime factorization of $2^n - 1$.

Problem 8.50

Read Problem 6.30 and prove, for $n \ge 2$ that if $n^m - 1$ is prime then $n = 2$.

CHAPTER 9

Counting Things

There is no problem in all mathematics that cannot be solved by direct counting.
- *Ernst Mach [1898]*

One of the first encounters a child has with mathematics is counting. This would seem to suggest that counting is the simplest thing one could possibly do within the borders of mathematics. The discovery of almost arbitrarily hard, and also of exceedingly annoying, counting problems caused a certain cognitive dissonance in mathematics which was solved by coming up with the term **combinatorics** for the more difficult tasks of counting. This both acknowledges the great depths and heights to which answering the question "how many?" can reach and permits counting to retain its childlike innocence.

This section rests firmly on the principles covered in Chapter 1 and so a quick review is an excellent idea at this point.

When counting the number of choices:

(i) independent choices multiply and

(ii) mutually exclusive choices add.

9.1 COMBINATORIAL PROOFS AND BINOMIAL COEFFICIENTS

This section introduces **combinatorial proofs**, a proof technique which is useful for establishing the validity of mathematical formula and equations. The binomial theorem may already be familiar to you from high school; it describes the algebraic expansion of powers of a binomial.

Theorem 9.1   Binomial Theorem

$$(x + y)^n = \sum_{k=0}^{n} \binom{n}{k} x^k y^{n-k}$$

The only symbol you many not have encountered is $\binom{n}{k}$, the **binomial coefficient**. We will give a *combinatorial* proof of the binomial theorem later in the section, but we begin by declaring what the binomial coefficient counts.

Definition 9.2   The number of subsets of size k in a set of size n is

$$\binom{n}{k}$$

which is spoken "n choose k." This number is the binomial coefficient with parameters n and k and is also sometimes written $C(n,k)$.

This symbol $\binom{n}{k}$ is both useful and causes many students to draw (and then rapidly erase) the horizontal lines that appear in fractions. We require a way of computing its value.

Proposition 9.3

$$\binom{n}{k} = \frac{n!}{k!(n-k)!}$$

Proof:

Suppose that we have n objects and are choosing a subset consisting of k of them. Then there are n objects available to be the first object. Having chosen the first object we may freely (independently) choose any of the $n-1$ remaining objects to be the second member of the set. There are, similarly $n-2$ choices for the third member, and, finally, $n-k+1$ choices for the kth element. We thus multiply these independent choices and count

$$n \cdot (n-1) \cdot (n-2) \cdots \cdots (n-k+1) = \frac{n!}{(n-k)!}$$

collections of k elements chosen from the set of size n. These collections are chosen in a particular order. A *subset* is unordered. Proposition 7.36 tells us that there are $k!$ permutations of a set of k elements. A permutation of a set can be thought of as a particular ordering of a set and so we see that the $n!/(n-k)!$ collections contain $k!$ copies of each unordered subset of k objects. Dividing by $k!$ we find that the number of k-element subsets of an n-element set is

$$\frac{n!}{k!(n-k)!}$$

giving the desired result.   □

The above is an example of a **combinatorial proof**. Combinatorial proofs are performed by counting the number of ways things could happen. Combinatorial proofs are a method of proof which cannot always be used (if, for instance, it is not actually possible or desirable to count the mathematical objects in the claim that is to be proved). However, they can be incredibly useful for practical mathematics, because they can be used to demonstrate the validity of equations and calcuations. Combinatorial proofs need to be combined with the other major proof techniques (the above proof was a direct proof).

Counting the number of ways to make an unordered set of k objects from a set of n objects has, as an intermediate step, counting the number of *ordered* choices of k objects. We define the name of these ordered lists and pull this step out as a corollary.

Definition 9.4   An ordered choice of k members of a set of n objects is called a **permutation of n objects taken k at a time** and the number of such ordered choices is denoted by $P(n, k)$ or alternatively as nPk.

Corollary 9.5   There are

$$P(n, k) = \frac{n!}{(n - k)!}$$

ordered choices of k objects from a set of n objects.

Proof:

This result appears in the proof of Proposition 9.3. □

Definition 9.6   Let A be a set of symbols. A **string** of length n over the *alphabet A*. Is a concatenation of n symbols from A. The length of a string is the number of symbols which appear in the string. The **empty string**, denoted λ, is the unique string of length 0.

Strings are a useful way to represent ordered information and are often used in computer science.

Example 9.7
The string 0112 is a string of length 4 taken from the alphabet $\{0, 1, 2\}$. The string $xxyyyx$ is a string of length 6 taken from the alphabet $\{x, y\}$.

At this point we know enough to prove the binomial theorem, but we start with an example which will make the proof easier to follow. It uses the distributive law aggressively but saves the commutative law until the last step.

Example 9.8

$$(x + y)^3 = (x + y)(x + y)(x + y)$$

$$(x + y)^3 = (xx + xy + yx + yy)(x + y)$$

$$(x + y)^3 = (xxx + xyx + yxx + yyx + xxy + xyy + yxy + yyy)$$

$$(x + y)^3 = x^3 + 3x^2y + 3xy^2 + y^3$$

Notice that the above directly verifies the binomial theorem for $n = 3$.

Combinatorial proof of the binomial theorem:

Notice that if we take $(x + y)^n$ then there are n terms in the multiplication, each of the form $(x + y)$. When we distribute these terms, but do not use the commutative law, we get 2^n terms, one for each possible choice of x or y in each multiplied term. Note that the 2^n terms are each strings of length n made up of xs and ys. When we collect these terms, the terms of the form $x^k y^{n-k}$ are the ones where we choose k xs from a total of n of the multiplied terms (the length n strings of x and y). There are $\binom{n}{k}$ ways to choose x in k of the n terms of an n-length string and the theorem follows. $\square$

The combinatorial proof of the binomial theorem directly shows how the binomial coefficients are involved. We now move on to investigate properties of the binomial coefficients.

Proposition 9.9

$$\binom{n}{k} = \binom{n}{n - k}$$

Proof:

This proof is left as an exercise.

Proposition 9.10   If $n \geq k \geq 1$ then

$$\binom{n}{k} = \binom{n - 1}{k} + \binom{n - 1}{k - 1}$$

Proof:

Let S be a set of n elements and let $x \in S$. Consider the set T of all k-element subsets of S. Then each member of T either contains x or fails to contain x. These options are mutually exclusive. The members of T that do not contains x are k elements chosen from the set $S - \{x\}$ which has $n - 1$ elements so there are $\binom{n-1}{k}$ such members of T. The remainder of the set in T, those that contain x, are x together with $k - 1$ points chosen from $S - \{x\}$. There are therefore $\binom{n-1}{k-1}$ such members of T. Since mutually exclusive choices add we see that

$$|T| = \binom{n}{k} = \binom{n-1}{k} + \binom{n-1}{k-1}$$

and the proposition follows. □

Proposition 9.10 shows that binomial coefficients are sums of other binomial coefficients. Blaise Pascal noticed that binomial coefficients have this property and came up with an ingenious method of diagramming them called **Pascal's triangle**. A portion of Pascal's triangle is shown in Figure 9.1. A small amount of additional information is required to build Pascal's triangle.

Proposition 9.11

$$\binom{n}{0} = \binom{n}{n} = 1$$

Proof:

There is one way to choose every member of a set—you choose the entire set. Proposition 9.9 tells us that there is therefore also one way to choose nothing at all. □

Definition 9.12 **Pascal's triangle** is the infinite diagram:

$$\binom{0}{0}$$
$$\binom{1}{0} \qquad \binom{1}{1}$$
$$\binom{2}{0} \qquad \binom{2}{1} \qquad \binom{2}{2}$$
$$\binom{3}{0} \qquad \binom{3}{1} \qquad \binom{3}{2} \qquad \binom{3}{3}$$
$$\cdots$$

Notice that each row of Pascal's triangle begins and ends with one and that other entries are computed with Proposition 9.10 by summing flanking entries above the entry you are currently computing.

Example 9.13

The binomial coefficients that appear in a row of Pascal's triangle are exactly those that appear in a binomial expansion $(x + y)^n$ with the row starting $1, n, \cdots$ being the relevant row. So:

$$(x + y)^5 = x^5 + 5x^4 y + 10x^3 y^2 + 10x^2 y^3 + 5xy^4 + y^5$$

$$
\begin{array}{ccccccccccc}
 & & & & & 1 & & & & & \\
 & & & & 1 & & 1 & & & & \\
 & & & 1 & & 2 & & 1 & & & \\
 & & 1 & & 3 & & 3 & & 1 & & \\
 & 1 & & 4 & & 6 & & 4 & & 1 & \\
1 & & 5 & & 10 & & 10 & & 5 & & 1 \\
\end{array}
$$

$$\cdots$$

Figure 9.1: The first six rows of Pascal's triangle.

Mild cleverness can extend this form of rapid computation from Pascal's triangle. Suppose we wish to compute $(x + 2)^5$. Separately computing the powers of 2 permit us to multiply the entries of two tables to get the desired coefficients:

Term	x^5	x^4	x^3	x^2	x	1
Binomial	1	5	10	10	5	1
Power-2	1	2	4	8	16	32
Product	x^5	$10x^4$	$40x^3$	$80x^2$	$80x$	32

So the answer is

$$x^5 + 10x^4 + 40x^3 + 80x^2 + 80x + 32$$

Up until this point, in mathematics, we usually begin counting at one. There are times when it is convenient to begin counting at zero and the numbering of the rows of Pascal's triangle is an instance of this phenomenon. In fact, in a lot of more advanced mathematical counting it makes sense to actually start counting at zero (mainly to correspond with the natural numbers which also start at zero).

The nth row of Pascal's triangle is the one that starts $1, n, \cdots$. This makes the topmost row the zeroth row. The following proposition has two different natural proofs, so we give both. The first is a combinatorial proof while the second exploits the binomial theorem.

Proposition 9.14   The sum of the nth row of Pascal's triangle is 2^n. Formally:

$$\sum_{k=0}^{n} \binom{n}{k} = 2^n$$

Proof 1:

Since $\binom{n}{k}$ is the number of sets with k elements that can be selected from an n-elements set, it follows that the sum counts all the subsets of an n-element set which Proposition 6.4 tells us is the number 2^n. □

Proof 2:

Starting with the binomial theorem, let x=y=1 and simplify:

$$(x + y)^n = \sum_{k=0}^{n} \binom{n}{k} x^k y^{n-k}$$

$$(1 + 1)^n = \sum_{k=0}^{n} \binom{n}{k} 1^k 1^{n-k}$$

$$2^n = \sum_{k=0}^{n} \binom{n}{k}$$

and the proposition follows. □

We will now look at some applications of binomial coefficients. It is often possible to reconfigure a problem into something that involves a binomial coefficient. We will start by defining a handy environment.

Definition 9.15   A **binary string** of length n is a string (or "word") of length n over the binary alphabet $\mathcal{B} = \{0, 1\}$.

Example 9.16
The binary strings of length 3 are 000, 001, 010, 011, 100, 101, 110, and 111.

Proposition 9.17   The number of binary strings of length n with exactly k zeros is $\binom{n}{k}$.

Proof 1:

We must *choose* k of the n positions for the zeros. □

While the above proof is correct, it is dangerously close to simply stating the claim. So we include a second, seperate proof which is hopefully helpful for students still trying to grasp the intuition required for this strange way of counting.

Proof 2:

Let S be a set with n elements. Number the elements from 1 to n. Consider each binary string as corresponding to a subset of S where the i-th position of the binary string is 0 if the i-th element is a member of the subset and 1 if it is not. In this manner, each binary string of length n with k zeros corresponds to exactly one subset of S with k elements. Since the set S has $\binom{n}{k}$ subsets of size k it follows that there are $\binom{n}{k}$ binary strings with exactly k zeros. □

Now some students may prefer Proof 1 over Proof 2 because it is shorter, and if you have a firm grasp of how the counting principles we've introduced work then it does make sense. However, the second proof is probably easier to grasp, and ties the binary strings to subsets explicitly. As you go on in mathematics it will become apparent that it's important to be able to adjust your proofs to the audience. Remember you are attempting to demonstrate a claim by means of a valid argument, but the important part is that your audience can actually follow along and see (and agree with) each step in the argument.

Suppose that we have k bins, numbered from 1 to k and a collection of n identical balls. How many different ways are there to put all the balls into the bins? We don't worry about the order that the balls are placed, just the number of balls in each bin when we are done. We start with an example.

Example 9.18

Suppose that we want to put two balls into three bins. Then either two balls both go in the same bin, which can happen three ways, or two bins get one ball and the third gets none. A picture of these possibilities is:

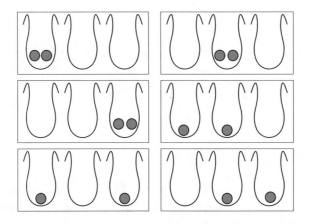

and we conclude there are six ways to put two balls into three bins.

Examine the first of the six configurations in Example 9.18. Replace balls with zeros and the spaces between bins with ones. Then you get the binary string 0011. In fact, all six of the configurations correspond to binary strings. If we present the corresponding strings in the same order we get:

$$0011 \quad 1001$$
$$1100 \quad 0101$$
$$0110 \quad 1010$$

and these are all $\binom{4}{2} = 6$ possible binary strings of length 4 with two ones. This suggests the solution to the balls-in-bins problem.

Proposition 9.19   There are

$$\binom{k + n - 1}{k}$$

ways to place k identical balls in n distinguishable bins.

Proof:

Examine the following correspondence. Place the bins side by side and look at any placement of balls. If the spaces between bins are mapped to ones and the balls to zeros, reading the entire configuration from left to right, then the correspondence is a function from ways to place balls in bins to the set of binary strings with $n - 1$ ones and k zeros. Any binary string of this sort specifies a particular arrangement of balls in bins as well so it is easy to see that this function is a bijection. This tells us that the number of ways to place balls in bins is equal to the number of binary strings with $n - 1$ ones and k zeros. The length of such a string is $k + n - 1$ and there are k zeros so we may apply Proposition 9.17 to calculate that there are

$$\binom{k + n - 1}{k}$$

ways to place the balls in the bins.   □

The proof of Proposition 9.19 uses an important technique. We located a bijection of ways to throw balls in bins with a particular sort of binary string. The binary strings were easy to count with a binomial coefficient and bijections preserve size. These facts gave us an easy way to count the number of configuration of balls in bins.

9.2   POKER HANDS AND MULTINOMIAL COEFFICIENTS

Another standard application of binomial coefficients is counting the number of ways a given poker hand can happen. We will be dealing with five-card hands dealt from a deck of 52-cards. These cards are evenly divided into the four suites clubs, diamonds, hearts, and spades. There are 13-cards of each suite with the values 2, 3, 4, 5, 6, 7, 8, 9, 10, Jack, Queen, King, Ace. When we have runs of cards that must be consecutive, like 4,5,6,7,8 the Ace is considered to both come before 2 and after King. It does not, however, wrap around, so King, Ace, 2, 3, 4 is *not* a run of five-cards.

Definition 9.20   The hand **two-of-a-kind** is a hand in which two-cards have the same value and the other three-cards all have different values, both from the first two and one another.

Example 9.21
Count the number of five-card hands that are two-of-a-kind.

Solution:

List the various independent choices that are required to get a two-of-a-kind hand and the number of ways they can be made:

Choice	Ways
Pick a value for the two-cards that are the same	13
Pick 3 values for the other three-cards	$\binom{12}{3}$
Pick the suits of the two-cards that are the same	$\binom{4}{2}$
Pick a suit for each of the three remaining cards	4^3

Multiplying these independent choices gives us

$$13 \cdot \binom{12}{3} \cdot \binom{4}{2} \cdot 4^3 =$$

$$13 \cdot 220 \cdot 6 \cdot 64 =$$

$$1{,}098{,}240$$

hands that are two-of-a-kind.

Definition 9.22 The hand **three-of-a-kind** is a hand in which three-cards have the same value and the other three-cards all have different values, both from the first three and one another.

Example 9.23

Count the number of five-card hands that are three-of-a-kind.

Solution:

List the various independent choices that are required to get a three-of-a-kind hand and the number of ways they can be made:

Choice	Ways
Pick a value for the three-cards that are the same	13
Pick 2 values for the other two-cards	$\binom{12}{2}$
Pick the suits of the three-cards that are the same	$\binom{4}{3}$
Pick a suit for each of the two remaining cards	4^2

Multiplying these independent choices gives us

$$13 \cdot \binom{12}{2} \cdot \binom{4}{3} \cdot 4^2 =$$

$$13 \cdot 66 \cdot 4 \cdot 16 =$$

$$54{,}912$$

hands that are three-of-a-kind.

Binomial coefficients can be thought of as choosing k objects from n or, equivalently, as dividing n objects into two sets, the chosen and unchosen. If we want to divide n objects into more than two categories there is a generalization of the binomial coefficient that can help us.

Definition 9.24 Suppose that we wish to break a set S with n elements into m disjoint sets

$$S_1, S_2, \ldots, S_m$$

with $|S_i| = n_i$. The number of ways this can be cone is the **multinomial coefficient**

$$\binom{n}{n_1\, n_1\, \cdots\, n_m}$$

When $m = 2$ the multinomial coefficient becomes the binomial coefficient. Notice that the fact the S_i are disjoint sets causes $n_1 + n_2 + \cdots + n_m = n$. With this in mind it is not too hard to prove that:

Proposition 9.25

$$\binom{n}{n_1 \; n_2 \; \cdots \; n_m} = \frac{n!}{n_1! n_2! \cdots n_m!}$$

Proof:

Adopt the setup of Definition 9.24. Pick an order for the elements of S; recall that there are $n!$ such orders. Let S_1 be the first n_1 elements of the order, S_2 be the next n_2 elements of the order, and so on. Then this partitions S into subsets of the desired size. If we re-order the elements of any S_i, however, we do not change the division of S into subsets. This means that $n!$ over-counts the multinomial coefficient by the product of the number of ways to reorder each subset - product because these re-orderings are independent. This product is $n_1! n_2! \cdots n_m!$ since there are $n_i!$ orderings of a set of size n_i. We divide $n!$ by the over-counting factor to obtain the desired formula for the multinomial coefficient.   □

Example 9.26

A class of 17 students is to be divided into 5 project teams of roughly equal size. The teams are allowed to choose their projects in the order that the groups are chosen and so it matters which team is chosen first. Since $17 = 3 \cdot 5 + 2$ we see that there are three groups of three and two groups of four. How many ways can this be done?

Solution:

Since the order that the groups are chosen matters, the three groups of three and two groups of four may be considered distinctly labeled disjoint sets. Thus, there are:

$$\binom{17}{3 \; 3 \; 3 \; 4 \; 4} = \frac{17!}{3! 3! 3! 4! 4!} = 2{,}858{,}856{,}000$$

(or about 2.86 billion) ways this can be done.

Something that is important to remember about binomial and multinomial coefficients is that, while the objects inside the groups chosen are not ordered the groups themselves *are* chosen in an order. If the groups in Example 9.26 had not been given the right to pick their projects in the order they were chosen the three groups of three students and the two groups of four students would have been interchangeable.

Interlude

The Birthday Conundrum

How many people do you need to have in a group before there is a 50% chance (or better) that two of the people have the same birthday?

This is a problem you may have encountered before, and we are going to use counting techniques to solve it here. Suppose we have a set S of n people and let $T = \{1, 2, \ldots, 365\}$ be the set of possible birthdays. Assume (falsely) that all birthdays are equally likely. We will count the number of ways to choose n distinct birthdays for n specific people and the number of functions

$$f : S \to T$$

that assign birthdays to people. When the ratio of these two numbers drops below 50% for the first time then the chance of two people having the same birthday is above 50%.

The number of functions $f : S \to T$ is 365^n because we make an independent choice of a birthday for each of the n people in S. Choosing n independent birthdays can be done $\binom{365}{n}$ ways. We must then assign a birthday to each of the n people, this can be done $n!$ ways. This means that the fraction of ways in which n people all have different birthdays is

$$\frac{\binom{365}{n} n!}{365^n}$$

Make a table to find when this number drops below 0.5.

People	Fraction	People	Fraction
1	1	13	0.805590
2	0.99726	14	0.776897
3	0.991796	15	0.747099
4	0.983644	16	0.716396
5	0.972864	17	0.684992
6	0.959538	18	0.653089
7	0.943764	19	0.620881
8	0.925665	20	0.588562
9	0.905376	21	0.556312
10	0.883052	22	0.524305
11	0.858859	**23**	**0.492703**
12	0.832975	24	0.461656

And we conclude that a group of 23 people is the smallest group in which we have better than 50% odds of the same birthday. In these calculations we assumed all birthdays are equally likely and also neglected leap years.

This idea may be easier to understand for the binomial coefficient. We know that $\binom{5}{3} = \binom{5}{2}$ but the first counts the number of ways to choose three of five and not choose two of five. The coefficient $\binom{6}{3} = 20$ but there are only ten ways to divide a set of six objects in half. In this latter case instead of "Choose a,b,c, do not choose d,e,f" or "Choose d,e,f and do not choose a,b,c" we instead have "The two groups are a,b,c and d,e,f" with no designation of chosen or not chosen. This significance of being chosen or not chosen (or of which distinct group you are assigned to) comes from the fact that the variables in the binomial theorem $(x + y)^n$ (or its generalizations) are different from one another.

Theorem 9.27   Trinomial theorem

$$(x + y + z)^n = \sum_{n_1+n_2+n_3=n,\ n_i \geq 0} \binom{n}{n_1\ n_2\ n_3} x^{n_1} y^{n_2} z^{n_3}$$

Proof:

This proof is left as an exercise.

Example 9.28   Trinomial expansion

$$(x + y + z)^3 = x^3 + y^3 + z^3 + 3x^2 y + 3x^2 z + 3xy^2 +$$

$$3xz^2 + 3y^2 z + 3yz^2 + 6xyz$$

Notice that there is considerable symmetry in the above expansion. Even through there are ten terms, only three different values of the trinomial coefficient appear.

The balls-in-bins problem required a bit of a leap (noticing the relevant bijection) in order to connect the problem with binomial coefficients. A few of the homework problems may be difficult because they require leaps of similar size. These are problems where you must exercise your imagination to find the solution: good luck.

9.3   THE INCLUSION-EXCLUSION PRINCIPLE

The Inclusion-Exclusion Principle is a fairly complex counting technique that can be motivated with simple examples. For the first example we need one piece of notation.

Definition 9.29 $\lfloor x \rfloor$ is the largest integer smaller than or equal to x while $\lceil x \rceil$ is the smallest integer larger than or equal to x.

Example 9.30
Count the integers in the range $1 \ldots 10$ (inclusive) that are not multiples of 2 or 3.

Solution:

In the range $1 \ldots 10$ there are: $\lfloor \frac{10}{2} \rfloor = 5$ multiples of 2, and $\lfloor \frac{10}{3} \rfloor = 3$ multiples of 3.

If we naively start with 10 and take away the numbers we don't want we will have: $10 - 5 - 3 = 2$ which is wrong. We can examine the 10 numbers and see that $1, 5$, and 7 are the numbers between 1 and 10 that are not multiples of 2 or 3. The key is that when subtracting the multiples of 2 and multiples of 3 we subtracted the multiple of $2 \cdot 3$ twice. So the way around that is to add the 1 multiple pf 6 back in. Thus, there are $10 - 5 - 3 + 1 = 3$ integers in the range $1 \ldots 10$ that are not multiples of 2 or 3.

Example 9.30 is the simplest example of a technique called *The Inclusion-Exclusion Principle*. However, it is so simple that it does not illustrate the entire idea. It is included as a simple example to digest before we move on to a more complex example.

Example 9.31
Count the integers in the range $1 \ldots 100$ that are not multiples of 2, 3, or 5.

Solution:

In the range $1 \ldots 100$ there are: $\lfloor \frac{100}{2} \rfloor = 50$ multiples of 2, $\lfloor \frac{100}{3} \rfloor = 33$ multiples of 3, and $\lfloor \frac{100}{5} \rfloor = 20$ multiples of 5.

Naively, we take 100 and take away the number we don't want giving us

$$100 - 50 - 33 - 20 = -3$$

which is silly. The problem is that if we add up all these multiples of the numbers we don't want to be multiple of we get 103 numbers. How is this possible? Well, we counted the multiples of 6 twice, both as multiples of 2 and as multiples of three. Likewise, we double-counted the multiples of 10 and 15.

There are: $\lfloor \frac{100}{6} \rfloor = 16$ multiples of 6, $\lfloor \frac{100}{10} \rfloor = 10$ multiples of 10, and $\lfloor \frac{100}{15} \rfloor = 6$ multiples of 15.

So if we compute

$$100 - 50 - 33 - 20 + 16 + 10 + 6 = 29$$

we have a plausible answer that is still wrong. The reason is that initially we counted, and so subtracted, the multiples of $2 \times 3 \times 5 = 30$ three times. When we added in the multiples of 6, 10, and 15 to compensate for subtracting them twice, we added the multiples of 30 back in three times. Since they were there originally, we are still counting them, and so we still need to subtract them. There are $\lfloor \frac{100}{30} \rfloor = 3$ multiples of 30. Subtracting those again, we see that there are

$$100 - 50 - 33 - 20 + 16 + 10 + 6 - 3 = 26$$

numbers in the range $1 \ldots 100$ that are not multiples of 2, 3, or 5.

The next step is to try to formalize the somewhat haphazard technique that organically solved the problems posed in Examples 9.30 and 9.31. We start by defining a traditional language for encoding things like "multiples of 2."

Definition 9.32   For a set S a **property** p on S is a quality that an element of S can have or not have. We denote by S_p the subset of elements of S that have property p.

Theorem 9.33   The Inclusion-Exclusion Principle
Let S be a set and let $P = \{p_1, p_2, \ldots, p_k\}$ be a set of properties on S. For $X \subseteq \{1, 2, \ldots k\}$ let

$$S_X = \bigcap_{i \in X} S_{p_i}$$

Then the number of elements of S that have none of the properties p_i is

$$\sum_{X \subseteq \{1,2,\ldots k\}} (-1)^{|X|} |S_X|$$

Proof:

This proof closely follows, in a more general form, the reasoning in Example 9.31. Notice that when $X = \emptyset$ we need to know what the intersection over an empty set is; it is the universal set and so equal to S in this case. When $|X| = 1$ we are subtracting elements with one (or more) properties. When $|X| = 2$ we are adding back the elements with two (or more) properties that we subtracted twice when $|X| = 1$. In general, when $|X| = k$, we are adding or subtracting elements with k or more of the properties, correcting for a previous over-or-under counting.

Consider elements that have exactly m properties. They are present initially, subtracted m times for various X of size 1, added in $\binom{m}{2}$ times for various X of size two. They are then subtracted $\binom{m}{3}$ times for various X of size 3. In general the contribution for elements of size $m > 0$, by the end of the sum, is

$$\sum_{i=0}^{m} (-1)^i \binom{m}{i} = \sum_{i=0}^{m} (-1)^i 1^{m-i} \binom{m}{i} = (1-1)^m = 0$$

This means the only elements of S that contribute to the sum are those that have zero properties.
□

The next example is similar to Example 9.31 but uses the formalism of the Inclusion-Exclusion Principle. Follow through this example in the context of the proof of Theorem 9.33.

Example 9.34
Calculate how many numbers in the range $1\ldots 1{,}000$ are not multiples of 3, 5, or 7. The properties are $p_1 = $ "is a multiple of 3;" $p_2 = $ "is a multiple of 5;" $p_3 = $ "is a multiple of 7." For the index set $\{1, 2, 3\}$ on these properties we have the subsets:

Subset	Meaning	Size
$\emptyset$	any number	1000
$\{1\}$	3n	333
$\{2\}$	5n	200
$\{3\}$	7n	142
$\{1, 2\}$	15n	66
$\{1, 3\}$	21n	47
$\{2, 3\}$	35n	28
$\{1, 2, 3\}$	105n	9

Applying the Inclusion-Exclusion Principle we get that there are

$$1{,}000 - 333 - 200 - 142 + 66 + 47 + 28 - 9 = 457$$

numbers that are not multiples of 3, 5, or 7.

Problem 7.63 asked for an estimate of the number of permutations without any fixed point. There is a name for this type of permutation.

Definition 9.35 A **derangement** is a permutation that has no fixed points.

Counting derangements is a natural target for the Inclusion-Exclusion Principle. We simply need to find a way to phrase fixed points as properties.

Proposition 9.36 The number of permutations on an n element set that fix at least m specific points is $(n - m)!$.

Proof:

Consider the one-line notation for permutations of the n element set $\{1, \ldots, n\}$. Given m specific points we will count the number of permutations which fix those points. Since the m points are

fixed the other $n - m$ points may be permuted in any order. There are $(n - m)!$ ways to do this.
□

Example 9.37
Find a formula for the number of derangements of a set of n points.

Consider the set S of all permutation of the set $\{1, 2, \ldots, n\}$. For $1 \le i \le n$ let p_i be the property that a permutation fixes the point i. If X is a set of fixed points from $\{1, 2, \ldots, n\}$ then Proposition 9.36 tells us there are $(n - |X|)!$ permutations that have all the properties p_i for $i \in X$. There are $\binom{n}{|X|}$ ways to choose a set X with a given number of fixed points. The variable $|X|$ has range $0 \ldots n$. Applying the Inclusion-Exclusion Principle there are

$$D_n = \sum_{i=0}^{n} (-1)^i \binom{n}{i} (n - i)!$$

derangements of a set with n elements.

Tabulating the first few terms of this formula we get:

n	1	2	3	4	5	6	7
D_n	0	1	2	9	44	265	1854

Noting that $\binom{n}{i} = \frac{n!}{i! \cdot (n-i)!}$ we see that the formula can be simplified.

$$
\begin{aligned}
D_n &= \sum_{i=0}^{n} (-1)^i \frac{n!}{i! \cdot (n - i)!} (n - i)! \\
&= \sum_{i=0}^{n} (-1)^i \frac{n!}{i!} \\
&= n! \sum_{i=0}^{n} \frac{(-1)^i}{i!}
\end{aligned}
$$

9.4   COUNTING SURJECTIONS

Recall that a function $f : X \to Y$ is a *surjection* if every element of Y is the image under f of some point in X. Another application of the Inclusion-Exclusion Principle is to compute the number of surjections of and n element set onto an m element set for $m \le n$. The key step is to define the properties we wish to avoid. If we want to hit every point in a set then the properties to avoid are those in which a function misses a given point.

Proposition 9.38   The number of functions from a set with n elements to a set with m elements is

$$m^n$$

Proof:

Let f be an arbitrary function between the two specified sets. For each element x in the n element set a value y may be independently chosen from the m element set to be $f(x)$. This means we make n independent choices when constructing f and each choice has m possible options for the selections. So from our work on simple counting in Chapter 1 we see that there are m^n ways to build f. □

Proposition 9.39   Suppose that $m \leq n$ are positive integers. The number of surjections of an n element set on to an m element set is

$$Surj(n,m) = \sum_{k=0}^{m}(-1)^k \binom{m}{k}(m-k)^n$$

Proof:

We will use the Inclusion-Exclusion Principle. Consider the set of all functions from an n-element set to an m-element set $\{1, 2, \ldots m\}$. Let property p_i be that a function maps no point to i. Then the surjections are those functions that possess none of these properties. For $X \subseteq \{1, 2, \ldots, m\}$, where X is a set of k elements, the functions that miss all the points in X are functions on to the remaining $m - k$ points. Proposition 9.38 tells us there are $(m - k)^n$ such functions. There are also $\binom{m}{k}$ ways to choose a k element set X. With all this in mind, the Inclusion-Exclusion Principle tells us there are

$$\sum_{k=0}^{m}(-1)^k \binom{m}{k}(m-k)^n$$

surjections. □

Since $Surj(n,m)$ is only non-zero for $1 \leq m \leq n$ it is possible to build a triangle, like Pascal's triangle, for $Surj(n,m)$. This triangle is shown in Figure 9.2.

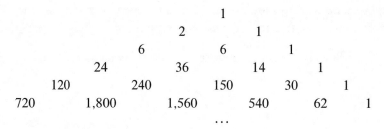

Figure 9.2: The first six rows of $Surj(n,m)$.

We have been through several different examples of applications of the Inclusion-Exclusion Principle. In all of them the key step is choosing a collection of properties that permit us to count the objects we are interested in. Each individual property specifies a subset of the set of objects we are working with and the counting finds the size of the complement of the union of all the sets.

9.5   ADDITIONAL EXAMPLES

Example 9.40
Suppose we have a graduating class with 75 students. How many different ways are there to:

(i) Pick a five-member homecoming committee?

(ii) Pick a Valedictorian and Salutatorian?

Solution:

(i) There are $\binom{75}{5} = 17,259,390$ ways to pick a 5-element subset from a set of size 75. Thus, there are 17,259,390 ways to select a 5-member homecoming committee from a class of 75 students.

(ii) The Valedictorian is the highest ranked student in the graduating class. Thus, there are 75 ways to pick a Valedictorian. The Salutatorian is the second highest ranked student in the graduating class. Thus, there are $75 - 1 = 74$ ways to pick the Salutatorian. Independent choices multiply, so there are $75 \cdot 74 = 5,550$ ways to pick a pair of Valedictorian and Salutatorian.

Example 9.41
Prove that

$$\sum_{k=0}^{n} \binom{n}{k} 2^k = 3^n$$

Proof:

By the binomial theorem, we have that:

$$(2+1)^n = \sum_{k=0}^{n} \binom{n}{k} 2^k 1^{n-k}$$

$$\Rightarrow (3)^n = \sum_{k=0}^{n} \binom{n}{k} 2^k$$

$\square$

Example 9.42
A **straight** is a 5-card poker hand in which the 5 cards have consecutive values but are not all the same suit. If the suits are the same this is a better hand, the **straight flush**. Compute the number of straights.

Solution:

First, we recall that when dealing with consecutive values the Ace may be used before a 2 or after a King. Next, we determine the number of ways to select 5 consecutive numbers from a standard deck of playing cards. If we use the smallest card value in the straight to specify the 5 consecutive values we see there are 10 ways (Ace through 10). Then, we determine the number of ways of assigning a suit to each of the 5 ordered cards. Independent choices multiply so there are 4^5 ways to do this. Finally, we subtract all those suit assignments which assign the same suit to all five-cards, there are 4 of them. Thus, there are $10 \cdot (4^5 - 4) = 10{,}200$ possible straights.

Example 9.43
A town with 24 street intersections is given 12 statues. Five are copies of a comic statue of a flying pig and seven are copies of a very serious looking beaver. If each statue is to be placed at a different intersection, how many different ways are there to place the statues?

Solution:

This can be solved using a multinomial coefficient. The key is to recognize that we wish to divide the set of 24 intersections into 3 disjoint sets. A set of size 5 for the comic statues of the flying pigs, a set of size 7 for the serious looking beavers and a set of size $24 - 5 - 7 = 12$ for all the intersections without a statue. Thus, there are $\binom{24}{5\ 7\ 12} = \frac{24!}{5!7!12!} = 2,141,691,552$ ways to place the statues.

Example 9.44

Suppose that at a shady poker game one of the players pockets one of the Aces. How many three-of-a-kind hands are now possible for the other players?

Solution:

We already know the number of three-of-a-kind hands with all 4 Aces in the deck from Example 9.23, there are 54,912 such hands. Next we pick a suit, WOLOG say Spades, and determine the number of three-of-a-kind hands which include the Ace of Spades.

First, there are those hands where the Ace of Spades is one of the three-of-a-kind. There are $\binom{3}{2}$ choices for the other suits of the two other Aces. There are $\binom{12}{2}$ ways to pick the values of the other two-cards, and 4^2 ways to pick the suits of those two-cards. Thus, there are $\binom{3}{2} \cdot \binom{12}{2} \cdot 4^2 = 3,168$ hands where the Ace of Spades is one of the three-of-a-kind.

Next, we determine the number of three-of-a-kind hands where the Ace of Spades is one of the two other cards. There are 12 ways to pick the value of the non-Ace three-of-a-kind and $\binom{4}{3}$ ways to pick the suits. There are 11 ways to pick the value of the other non-three-of-a-kind card and 4 ways to pick its suit. Thus, there are $12 \cdot \binom{4}{3} \cdot 11 \cdot 4 = 2,112$ three-of-a-kind hands where the Ace of Spades is one of the two other cards.

Therefore, there are $54,912 - 3,168 - 2,112 = 49,632$ three-of-a-kind hands which do not include the Ace of Spades. Since it did not really matter which Ace we singled out, there are 49,632 three-of-a-kind hands still possible when one of the Aces is pocketed.

9.6   PROBLEMS

Problem 9.45
Prove Proposition 9.9.

Problem 9.46
Compute the next three rows of Pascal's triangle and add them to the bottom of the triangle given in Figure 9.1.

Problem 9.47
Suppose we have a club with 12 members. How many different ways are there to

  (i) choose a president, secretary, and treasurer?

  (ii) choose a three-member executive committee?

Problem 9.48
Prove that all the prime numbers in Pascal's triangle appear adjacent to a 1.

Problem 9.49
Find the smallest binomial coefficient with six distinct prime factors.

Problem 9.50
Prove:
$$\sum_{k=0}^{n} \binom{n}{k}(-1)^k = 0$$

Problem 9.51
Multiply out and simplify:

  (i) $(x + 1)^8$, (ii) $(x + 2)^6$, (iii) $(2x + 1)^6$, (iv) $(2x + 3y)^4$, (v) $(x - 3)^5$, and (vii) $(x^2 + 1)^7$.

Problem 9.52
Prove that
$$\sum_{k=1}^{n} k\binom{n}{k} = n \cdot 2^{n-1}$$
you may use calculus if you wish.

Problem 9.53
For the birthday conundrum, which of the two assumptions, that all birthdays are equally likely,

or that it is not a leap year, make the biggest difference in the estimation of the number of people needed to have a 50% chance of the same birthday?

Problem 9.54
Prove Proposition 9.14 by induction on n.

Problem 9.55
Compute the number of ways to place k indistinguishable balls in n bins numbered 1 to n if each bin gets at least one ball.

Problem 9.56
Suppose that we have k blue balls, m red balls, and n bins numbered $1, 2, \ldots, n$. How many ways are there to place the balls in the bins?

Hint: Place the blue balls first, then the red balls.

Problem 9.57
Prove that

$$\sum_{k=0}^{n} \binom{n}{k}^2 = \binom{2n}{n}$$

There are both direct and combinatorial proofs possible.

Problem 9.58
A five-card hand in which there are two pairs that have the same value, but values that are different from one another, and the fifth-card has a third value is called **two pair**. Compute the number of poker hands that are two pair.

Problem 9.59
A **full house** is a five-card poker hand in which there are three-cards of one value and two of another. Compute the number of hands that count as being a full house.

Problem 9.60
A **flush** is a five-card poker hand in which all five-cards have the same suit but do not have five consecutive values. Compute the number of flushes.

Problem 9.61

A **straight flush** is a five-card poker hand in which the cards are all the same suit and have five consecutive values. Compute the number of such hands.

Problem 9.62

A group of ten volunteers that are sprucing up a city park are divided into three groups whose jobs are collecting trash, repainting the picnic shelters, and cleaning and restoring the signs along the nature walk. Repainting takes four people and the other two groups need three people. How many ways are there to divide up the teams?

Problem 9.63

Suppose we are painting a sculpture consisting of a cube that is mounted with a corner on the ground so that it has two blue, two red, and two yellow sides. How many different designs are possible?

Problem 9.64

Prove Theorem 9.27.

Problem 9.65

A *permutation matrix* is an $n \times n$ matrix with a single one in each row and column and zeros everywhere else. An example is shown below. As a function of n, how many permutation matrices are there?

$$\begin{bmatrix} 0 & 1 & 0 & 0 & 0 \\ 1 & 0 & 0 & 0 & 0 \\ 0 & 0 & 1 & 0 & 0 \\ 0 & 0 & 0 & 0 & 1 \\ 0 & 0 & 0 & 1 & 0 \end{bmatrix}$$

Problem 9.66

Read Problem 9.65. How many ways are there to construct an $n \times n$ matrix in which n entries are one and the remaining entries are zero? What fraction of this type of $n \times n$ matrices are permutation matrices?

Problem 9.67

Suppose that a class has 22 students in the Math-stats club, 11 students in the Astronomy club, and 5 that are in both. How many students in this class, total, are in at least one of these clubs? In addition to computing the numerical answer give a Venn diagram that illustrates the situation.

Problem 9.68
Make a Venn diagram illustrating how the Inclusion-Exclusion Principle operates when counting the numbers m in the range $1 \leq m \leq 32$ that are not multiples of 2, 3, or 5.

Problem 9.69
Suppose that during an evacuation of a fancy restaurant a coat-check employee throws 12 coats toward their owners at random. Assuming the owners are not paying attention but that each catches a coat, what are the odds that no-one gets the right coat?

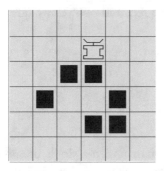

Problem 9.70
The *Tartarus* problem is a simple AI test problem. At the beginning of a case of the Tartarus problem, a bulldozer is placed on a 6×6 grid with six boxes that cannot be against the edge of the grid. An example is shown above. The bulldozer may face in any of the four directions that line up with the grid. If the boxes are indistinguishable from one another then how many initial conditions for the Tartarus problem are there? Assume that the bulldozer is also not permitted to be adjacent to the wall when it starts.

Problem 9.71
If D_n is the number of derangements of an n set, prove that: $D_n = n \cdot D_{n-1} + (-1)^n$.

Problem 9.72
Extend the table of value for D_n given in the text to $n = 10$.

Problem 9.73
Suppose that F_n is the number of permutations of the set $\{1, 2, \ldots n\}$ with exactly one fixed point. Prove that F_n and D_n differ by one.

Problem 9.74
Find formulas for $Surj(n, 1)$, $Surj(n, 2)$, and $Surj(n, n)$ that do not involve summations.

Problem 9.75
Proposition 9.10 encodes the rule for computing an entry in Pascal's triangle in terms of the two entries above it and to the left and right. Find the corresponding rule for the $Surj(n, m)$ triangle.

Problem 9.76
Prove that

$$n! = \sum_{k=0}^{m} (-1)^k \binom{n}{k} (n - k)^n$$

Problem 9.77
Suppose we have n competitors in a tournament. Then there are $n!$ possible rankings if there are no ties. Compute the number of rankings if ties, up to and including an n-way tie, are possible.

CHAPTER 10

Relations

In this chapter we will formally define relations. Some relations such as "less than or equal to" or "divides" are already familiar to you. The formal definition will permit us to explore properties of these relations and use them to create new structures.

Definition 10.1   A **Binary Relation** (or simply a **Relation**) R on a set S is a set of ordered pairs from $S \times S$. If $(a, b) \in R$ then we say that "a is related to b."

The most familiar relation is **equality**. The equality relation on a set S is the set of ordered pairs

$$\{(x, x) : x \in S\}$$

Two elements of S are related if and only if they are the same element. Relations are classified by the properties they have. Here are some of the properties a relation can have.

Definition 10.2   A relation R on a set S is **reflexive** if $(s, s) \in R$ for all $s \in S$.

Definition 10.3   A relation R is **symmetric** if whenever $(a, b) \in R$ then we also have that $(b, a) \in R$.

Definition 10.4   A relation is **anti-symmetric** if whenever both $(a, b) \in R$ and $(b, a) \in R$ we have that $a = b$.

Definition 10.5   A relation is **transitive** if whenever $(a, b) \in R$ and $(b, c) \in R$ then we also have that $(a, c) \in R$.

In order to be useful or interesting, relations need to have some structure. The set

$$R = \{(1, 2), (7, 9), (3, 4)\}$$

is a relation but it doesn't really have any cool properties and its hard to envision an application. The set of all relations on R is $\mathcal{P}(R \times R)$, the set of all subsets of the Cartesian product of R with itself. Notice there is an empty relation with no related pairs. There are some types of relations that show up again and again in mathematics that are very useful.

10.1   EQUIVALENCE RELATIONS AND PARTITIONS

Definition 10.6   An **equivalence relation** is any relation that is reflexive, symmetric, and transitive.

The following example is included as a template for proving a relation is an equivalence relation. Such a proof is usually performed by checking the three required properties. It is somewhat unsatisfying as all three steps use the definition of equality.

Definition 10.7   Two numbers are **equal** if their difference is zero.

Example 10.8
Show that the equality relation on any set S of numbers is an equivalence relation.

Solution:

All we have to do is check the three properties required for equality to be an equivalence relation. For every element of $s \in S$, $s - s = 0$ and so (s, s) is in the relation for all S and so equality is reflexive (every number is equal to itself). If $a - b = 0$ then we see that $-(a - b) = -0$ and so $b - a = 0$ which tells us that when $a = b$ it is also the case that $b = a$ and so equality is symmetric. If $a = b$ and $b = c$ the definition of equality ensures $a - b = 0$ and $b - c = 0$. Adding these equations yields $a - b + b - c = a - c = 0$ and we also have that equality is transitive.

Let's look at another example of an equivalence relation. Consider the space of convex polygons in the Euclidean plane. Define two polygons to be **translation-equivalent** if one can be translated (moved without being rotated at all) to cover exactly the same points in the plane as the other. Since two equivalent polygons can be in different locations this relation is not simply the equality relation. It is, however, an equivalence relation.

Proposition 10.9   Translation-equivalence (TE) is an equivalence relation.

Proof:

A polygon already covers the same points in the plane as itself after making a translation of no distance at all in the plane. This demonstrates that TE is reflexive. If we translate one polygon so that it covers the same points in the plane as another then a translation in exactly the opposite direction will permit us to cover the first polygon with the second. It follows that TE is symmetric. Suppose that polygons P and Q are translation equivalent by a translation t in the plane and the polygons Q and W are translation equivalent by a translation s. Then if we apply t and then s to P it will first translate P to exactly cover Q and then translate the points of P to exactly cover W. It follows that TE is transitive and so satisfies all the properties of an equivalence relation.   □

Equivalence relations are often used to simplify or reduce a structure. We will now build up an important example of this application of equivalence relations. The result is a new mathematical structure we will use for a number of purposes.

Definition 10.10   Let n be a positive natural number. Define a relation **equivalence modulo n** on the integers to consist of the ordered pairs (u, v) such that n divides $(u - v)$. If $n|(u - v)$ we say that u is congruent to v (*mod n*) or that u is equivalent to v (*mod n*).

Equivalence of x and y modulo n is denoted

$$x \cong_n y$$

Proposition 10.11   Equivalence (*mod n*) is an equivalence relation.

Proof:

Since $0 = n \cdot 0$ it follows that $n|0$ and so $n|(u - u)$ for any u. We may conclude that equivalence modulo n is reflexive. Suppose that $u \cong_n v$. Then $n|(u - v)$ and so for some k, $u - v = kn$. This implies that $v - u = (-k)n$ telling us that $n|(v - u)$ and hence $v \cong_n u$. It follows that the relation is symmetric. It remains to test for transitivity. Suppose that we have $x \cong_n y$ and $y \cong_n z$. By definition this means that $n|(x - y)$ and $n|(y - z)$, and so we may find k and q so that

$$\begin{aligned} x - y &= kn \\ y - z &= qn \end{aligned}$$

adding these two equations we obtain $x - z = (k + q)n$ and so $n|(x - z)$ and we see that $x \cong_n z$. This gives us transitivity and the proposition follows.   □

Definition 10.12   If R is an equivalence relation on a set S then for $x \in S$ we denote by $[x]$ the subset of S consisting of elements that are related to x. In curly brace notation

$$[x] = \{y \in S : (x, y) \in R\}$$

The subset $[x]$ is called the **equivalence class** of x.

If we are only discussing a single equivalence relation R then the meaning of the symbol $[x]$ for equivalence classes is unambiguous; if there is the potential for confusion we denote the equivalence class of x under the relation R by $[x]_R$. For equivalence modulo n we write $[x]_n$ for equivalence classes.

Proposition 10.13   If $y \in [x]$ then $[x] = [y]$.

Proof:

The proof is left as an exercise.

Equivalence classes are a useful way to discuss the **partition** of a set.

Definition 10.14   For any (non-empty) set S, a **partition** of S is a collection of non-empty pairwise disjoint subsets of S whose union is S. In other words, $\{A_i\}_{i \in B}$ (for some index set B) such that: for all $i \in B$, $A_i \subseteq S$, $A_i \neq \emptyset$, and $\cup_{i \in B} A_i = S$ and for $i \neq j, i, j \in B$ $A_i \cap A_j = \emptyset$. The collection of sets $\{A_i\}_{i \in B}$ is said to **partition** S. The A_i are called the **cells** of the partition.

Proposition 10.15   The equivalence classes of an equivalence relation R on a set S form a partition of S, where each equivalence class corresponds to a cell of the partition.

Proof:

Let S be a non-empty set, and let R be an equivalence relation on S such that $\{X_i\}_{i \in A}$ is the set of equivalence classes of S. By definition of an equivalence class, each $X_i \subseteq S$. Since S is non-empty, and R is reflexive, $(x, x) \in R$ for all $x \in S$ so there is at least one element in each equivalence class, X_i. Thus, $X_i \neq \emptyset$ for all $i \in A$, and since each element of S is in some equivalence class it follows that $\cup_{i \in A} X_i = S$. Now consider distinct equivalence classes X_i, X_j ($i \neq j$) if more than one equivalence class exists. If $a \in X_i$ then $X_i = \{b \in S : (a, b) \in R$ thus for any $c \in S$ s.t. $c \notin X_i$ $(a, c) \notin R$ and hence if $i, j \in A$ with $i \neq j$ then $X_i \cap X_j = \emptyset$. Thus, the set of equivalence classes $\{X_i\}_{i \in A}$ satisfies the definition of a partition of S. $\square$

Proposition 10.16   A partition of a non-empty set S into the collection $\{A_i\}_{i \in B}$ defines an equivalence relation R on S when xRy iff x, y are elements of the same cell A_i.

Proof:

This proof is left as an exercise.

Proposition 10.15 and it's converse Proposition 10.16 together demonstrate that equivalence relations and partitions are essentially alternate ways of describing the same collection of mathematical objects.

10.2 THE INTEGERS MODULO n

Now that we have the ideas of equivalence classes and partitions, we can use the equivalence relation modulo n to create a new structure.

Definition 10.17 The **integers modulo** n are a number system whose members are the equivalence classes of the relation $\cong_n$ on the integers.

The symbol $\mathbb{Z}_n$ is used to denote the integers modulo n. We still need to pin down how arithmetic works in $\mathbb{Z}_n$. We start by defining the addition and multiplication of equivalence classes.

Definition 10.18 The **sum** of two equivalence classes of integers $[x]_n$ and $[y_n]$ is the set

$$[x]_n + [y_n] = \{a + b : a \in [x]_n, b \in [y]_n\}$$

Definition 10.19 The **product** of two equivalence classes of integers $[x]_n$ and $[y_n]$ is the set

$$[x]_n[y_n] = \{ab : a \in [x]_n, b \in [y]_n\}$$

These definitions, while reasonable, are pretty mechanical. The following proposition shows why these are natural definitions in this case.

Proposition 10.20 **Consistency of** $\mathbb{Z}_n$

(i) $[x]_n + [y]_n = [x + y]_n$

(ii) $[x]_n[y]_n = [xy]_n$

Proof:

Suppose that $a \cong_n b$ and that $u \cong_n v$. Then for some k and m

$$
\begin{aligned}
a - b &= kn \\
u - v &= mn \\
(a + u) - (b + v) &= (k + m)n
\end{aligned}
$$

This tells us $a + u \cong_n b + v$. Since the pairs of numbers that were congruent (*mod n*) were chosen without constraint, this demonstrates that addition of sets preserves equivalence classes (*mod n*). Turning to multiplication,

$$
\begin{aligned}
(a - kn)(u - mn) &= bv \\
au - knu - amn + kmn^2 &= bv \\
au - bv &= (ku + am - kmn)n
\end{aligned}
$$

So $au \cong_n bv$, and we see similarly that multiplication of sets preserves equivalence classes (*mod n*). □

A side effect of the fact that equivalence (*mod n*) is consistent with integer addition and multiplication in the fashion demonstrated in Proposition 10.20 means that $\mathbb{Z}_n$ inherits many of the arithmetic properties of the integers such as associativity and commutativity of addition and multiplication and the distributive law.

For the sake of notational convenience instead of writing $ab \cong_n = c$ we will sometimes write $ab = c$, using $=$ instead of $\cong_n$ when it is clear that we are discussing the integers modulo n. This allows us to easy write inequality such as $\neq$ as well. Likewise, we will state k rather than $[k]_n$ when it is clear that we are discussing the equivalence class of k in the intergers modulo n.

There are some properties of the integers that $\mathbb{Z}_n$ does *not* inherit. Notice that the additive identity of $\mathbb{Z}_n$ is $[0]_n$, the set of all multiples of n and that the multiplicative identity is $[1]_n$. Suppose that $n = 4$ then since two times two equals four in the integers it follows that two times two equals zero (*mod 4*). This is a violation of a fact about the integers we have used before, that $ab = 0$ implies that $a = 0$ or $b = 0$.

Definition 10.21   Suppose $[a]_n[b]_n = [0_n$ and we have that $[a]_n \neq [0]_n$ and $[b]_n \neq [0]_n$ then we call $[a]_n$ and $[b]_n$ **zero divisors**. The equivalence class of zero itself is not a zero divisor. Informally, we say that a and b are zero divisors and zero itself is not a zero divisor.

With the above definition we can shorten the demonstration before the definition to the statement "two is a zero divisor in $\mathbb{Z}_4$." The existence of zero divisors is somewhat annoying, but there is an even more obnoxious property available. We first need to define what is meant by a natural number power.

Definition 10.22   Suppose $n \in \mathbb{N}$ and that a is an object for which a form of multiplication $\cdot$ is defined. Suppose also that the multiplication has an identity, 1. Then we define exponentiation by a natural number as follows:

   (i) $a^0 = 1$,

   (ii) $a^1 = a$, and

(iii) $a^n = a \cdot a^{n-1}$.

Notice that this notion of exponentiation by a natural number is exactly the one you are already used to.

Definition 10.23   Suppose that $a \neq 0$ and that for some finite natural number $n \geq 1$ we have $a^n = 0$. Then we say that a is **nilpotent**.

Notice that $([2]_4)^2 = [2]_4 \cdot [2]_4 = [0]_4$ so we see that two is nilpotent in $\mathbb{Z}_4$.

Definition 10.24   If $a, b \in \mathbb{Z}_n$ have the property that $ab = [1]_n$ then we say that a and b are **units** in $\mathbb{Z}_n$.

The proof of the next proposition gives us valuable practice with the notion that the members of $\mathbb{Z}_n$ are each infinite equivalence classes on $\mathbb{Z}$. Recall, for example that zero in $\mathbb{Z}_n$ is the set of all multiples of n.

Proposition 10.25   If $a \in \mathbb{Z}_n$ then either a is a unit, a zero divisor, or equal to zero.

Proof:

Suppose that $a \neq 0$. Recall that $a = [x]_n$ for an infinite number of different $x \in \mathbb{Z}$. By applying the division theorem to any such x we may find a unique integer $y \in a$ so that $0 < y < n$ and $a = [y]_n$. The fact that $y > 0$ follows from the hypothesis that $a \neq 0$.

Now, suppose that $GCD(y, n) = r > 1$. Then $1 < \frac{n}{r} < n$. Let $s = \frac{n}{r}$ and let $k = \frac{y}{r}$ notice that $y \cdot s = k \cdot r \cdot (\frac{n}{r}) = k \cdot n$. Then applying Proposition 10.20 demonstrates that $[kn]_n = [n]_n = [y]_n \cdot [s]_n$. This is equivalent (*mod n*) to $0 = a \cdot [s]_n$. But we also have $1 < s < n$ and so $[s]_n \neq 0$. It follows that a is a zero divisor.

If, on the other hand, $GCD(y, n) = 1$ then we know that 1 is a linear combination of y and n so that $1 = uy + vn$ for some $u, v \in \mathbb{Z}$. Taking this expression (*mod n*) and applying Proposition 10.20 we get $[1]_n = [u]_n [y]_n + [v]_n [n]_n$. Recalling that $[n]_n$ is zero in $\mathbb{Z}_n$ and that $[1]_n$ is one in $\mathbb{Z}_n$, this simplifies to $[1]_n = [u]_n \cdot a$ and we have that a is a unit. The proposition follows.   □

Example 10.26
Tables for $\mathbb{Z}_6$ The integers (*mod n*) are a finite set and so, for small n, it is possible to give complete addition and multiplication tables. The following are such tables for $\mathbb{Z}_6$.

Addition								**Multiplication**					
+	[0]	[1]	[2]	[3]	[4]	[5]	·	[0]	[1]	[2]	[3]	[4]	[5]
[0]	[0]	[1]	[2]	[3]	[4]	[5]	[0]	[0]	[0]	[0]	[0]	[0]	[0]
[1]	[1]	[2]	[3]	[4]	[5]	[0]	[1]	[0]	[1]	[2]	[3]	[4]	[5]
[2]	[2]	[3]	[4]	[5]	[0]	[1]	[2]	[0]	[2]	[4]	[0]	[2]	[4]
[3]	[3]	[4]	[5]	[0]	[1]	[2]	[3]	[0]	[3]	[0]	[3]	[0]	[3]
[4]	[4]	[5]	[0]	[1]	[2]	[3]	[4]	[0]	[4]	[2]	[0]	[4]	[2]
[5]	[5]	[0]	[1]	[2]	[3]	[4]	[5]	[0]	[5]	[4]	[3]	[2]	[1]

Notice that the tables make it easy to spot zero divisors and units.

Let's stop and take a look at the broad view. The integers $\mathbb{Z}$ are an infinite set. We defined an equivalence relation $\cong_n$ on the integers. Taking the equivalence classes with respect to this equivalence relation yielded a system of equivalence classes $\mathbb{Z}_n$ that retain some, but not all, properties of the integers. The various $\mathbb{Z}_n$ form an infinite number of arithmetic systems with a variety of interesting properties. The process we used to create the various $\mathbb{Z}_n$ from the integers is called *moding out by an equivalence relation* or *reducing by an equivalence relation*. The equivalence relation $\cong_n$ turns out to preserve a remarkable amount of structure. The amount of old structure preserved and the amount of interesting new structure that forms when we mod out by an equivalence relation varies substantially with the details of the system being reduced and the equivalence relation.

Examine again the equivalence relation on convex polygons given in this section. This equivalence relation creates infinite equivalence classes of polygons that are the same shape and in the same orientation but may be sitting at different position in the plane. Translation-equivalence abstracts out the important properties of shape and orientation, separating them from the property of position. Likewise, $\cong_n$ separates out the property "what is the remainder of a when divided by n" from the other properties a may have. All equivalence relations can be seen as abstracting out *some* property. Interesting or useful equivalence relations abstract out interesting or useful properties. We now move on to another important type of relation.

10.3   PARTIAL ORDERS

Definition 10.27   A **partial order** on a set S is a reflexive, anti-symmetric, transitive relation on S.

Partial orders are sometimes called, **partially ordered sets, order relations, ordered sets** or even **posets**. The normal ordering of the real numbers $x \leq y$ is an example of a partial order (sometimes called the **usual order**). Let us give a formal definition of $\leq$ that can be used in writing proofs.

Definition 10.28   For two real numbers $a, b \in \mathbb{R}$ we say that $a \leq b$ if for some non-negative real number c we have $a + c = b$.

The statement of the next proposition introduces a notation for partial orders in which the set and the symbol for the relation are presented as an ordered pair $(set, relation)$.

Proposition 10.29   The relation $(\mathbb{R}, \leq)$ is a partial order.

Proof:

Since each number is equal to itself, each number is less than or equal to itself and so we have that $\leq$ is reflexive. If $a \leq b$ and $b \leq a$ then we have for non-negative c and d that $a + c = b$ and $a = b + d$. This tells us that $b + d + c = b$ which in turn implies that $c + d = 0$. This tells us $c = -d$ which, given that c and d are non-negative, forces $c = d = 0$. From this we may deduce that $a = b$ and so have proven that $\leq$ is anti-symmetric. It remains to verify transitivity. Suppose $a \leq b$ and $b \leq c$. Then we may find non-negative x and y so that $a + x = b$ and $b + y = c$. Substitution yields $a + (x + y) = c$. We know that the sum of non-negative numbers is non-negative and so we have $a \leq c$, in other words transitivity. Since $\leq$ is reflexive, anti-symmetric, and transitive we see that it is a partial order. $\square$

Definition 10.30   If a partial order P on the set S has the additional property that for all $x, y \in S$ either $(x, y) \in P$ or $(y, x) \in P$ we say P is a **total order** (or a **linear order**).

Proposition 10.31   The relation $(\mathbb{R}, \leq)$ is a total order.

Proof:

By the previous proposition, we already know that $\leq$ partially orders $\mathbb{R}$. Suppose that $a, b \in \mathbb{R}$. By the trichotomy principle, $a - b$ is positive, negative, or zero. If $a - b = 0$ then $a = b$ and we have $a \leq b$ and $b \leq a$. If $c = a - b > 0$ then $a = b + c$ and we have $b \leq a$. If, on the other hand, $c = a - b < 0$ then $b = a + (-c)$ for $-c$ non-negative and we have $a \leq b$. In all three cases a and b are in the relation in some order. It follows that $\leq$ is a total order. $\square$

It's useful to have counter-examples as well as examples so we'll now look at a partial order which is not a total order: Set Inclusion.

Proposition 10.32   Let S be a non-empty set. The relation $(\mathcal{P}(S), \subseteq)$ is a partial order (that is not a total order when S contains at least two elements). This relation is known as **set inclusion**.

Proof:

We will first show that $\subseteq$ is a partial order. Let X, Y and Z be subsets of S. For any set X we have that $X \subseteq X$, since each element of X is an element of X, so $\subseteq$ is reflexive. If $X \subseteq Y$ and $Y \subseteq X$ we have that $X = Y$ by Proposition 5.23 (the first proposition about sets that we proved), so $\subseteq$ is anti-symmetric. If $X \subset Y$ and $Y \subseteq Z$ then for each $x \in X$ we have that $x \in Y$ and hence $x \in Z$ so $X \subseteq Z$ thus $\subseteq$ is transitive. Therefore, $(\mathcal{P}(S), \subseteq)$ is a partial order.

If there are at least two distinct elements $a, b \in S$ then $a \notin \{b\}$ and $b \notin \{a\}$ so the sets $\{a\}$ and $\{b\}$ are not related to each other in any way and it follows that $\subseteq$ is not a total order. $\qquad \square$

Another partial order you may never have considered is: divisibility. Notice that we define the relation not on the entire set of integers, rather only for those integers that are not negative. This choice is explained in the exercises.

Proposition 10.33   The divisibility relation on the non-negative integers is a partial order.

Proof:

This proof is an exercise.

Divisibility on the positive whole numbers is an interesting partial order with a great deal of structure. It is *not* a total order, though it contains some infinite subsets that are totally ordered by divisibility, as we will see in the exercises.

Definition 10.34   We say two elements a and b of a partial order R are **immediately related** if $(a, b) \in R$, $a \neq b$, and there exists no c so that (a, c) and (c, b) are in R.

The notion of *immediately related* captures the idea that two points are adjacent in the partial order—they are distinct, related to one another, and have nothing "between" them in the partial order.

Definition 10.35   The **Hasse diagram** of a partial order is a picture of the elements with links between all pairs of immediately related elements. The links may be shown as line segments or arrows denoting the direction of the relation. In the event that arrows are not used, vertical position is kept consistent with the ordering, as in the example below.

Example 10.36
Give the Hasse diagram of the set of positive divisors of 12 under divisibility.

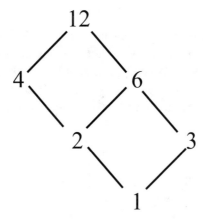

Notice that for any two numbers in the diagram if $a|b$ then a is connected to b by a chain of upward links. There are a number of pairs of numbers that are not related in either direction, e.g., 4 and 6. Neither divides the other, verifying that divisibility is not a total order.

Proposition 10.37   Suppose that $(X, \preceq)$ is a partial order on the set X and that $Y \subseteq X$. Let $(Y, \prec)$ be the subset of $(X, \preceq)$ consisting of those ordered pairs in which both members of the ordered pair are drawn from Y. Then $(Y, \prec)$ is a partial order on Y.

Proof:

This proof is left as an exercise.

Proposition 10.37 means that for each partial order $(S, \leq)$ that we define we gain a collection of partial orders that live, in a sense, inside that partial order. When a partial order is infinite it contains a huge collection of other partial order.

10.4   ADDITIONAL EXAMPLES

Example 10.38
Count the number of reflexive relations on a three element set.

Solution:

A relation is made up of ordered pairs. There are 3×3 possible ordered pairs. A reflexive relation must contain all ordered pairs of the form (x, x), so every reflexive relation must contain the 3 ordered pairs (a, a), (b, b), and (c, c). Thus, there are $3 \cdot 3 - 3 = 6$ ordered pairs which we can choose to include or not include in the reflexive relation, each of these binary choices is independent, therefore there are $2^6 = 64$ different reflexive relations on a three element set. $\square$

Note: The preceding solution contained perhaps a bit more information than was needed to explain each step, but it's important to at least be able to explain each step. *Where are the 2's coming from? Binary choices. Why are we multiplying them together rather than just adding them? Because they are independent choices not mutually exclusive choices. If someone asked you to explain a step, could you do so easily or would you have to stare at a calculation for a few minutes to figure out why you counted it that way?*

Example 10.39
Show that the relation $(\mathbb{R}, R)$ given by: aRb if and only if $a = b$ or $(a + b)^2 = (a + b)$ with $(a + b)^2 > 0$, is an equivalence relation.

Solution:

(Reflexivity) Since we have that $(a, b) \in R$ whenever $a = b$ directly as part of the definition we have reflexivity.

(Symmetry) Suppose $(a, b) \in R$ then either $a = b$ in which case $(b, a) \in R$ or $(a + b)^2 = (a + b)$, in which case the commutativity of addition of real numbers gives us that $(b + a)^2 = (b + a)$ and hence $(b, a) \in R$.

(Transitivity) Suppose (a, b), $(b, c) \in R$. If $a = b$ or $b = c$ then transitivity follows trivially, so assume $a \neq b$ and $b \neq c$. Then $(a + b)^2 = (a + b)$, this means that either $|a + b| = 1$ (in which case $(a + b) = 1$ since $(a + b)^2 \geq 0$) or $(a + b) = 0$. However, $(a + b)^2 > 0$ so $a + b \neq 0$. Similar results hold which imply that $b + c = 1$. If $(a + b) = 1$ and $b + c = 1$ then $a = 1 - b = c$ and hence $(a, c) \in R$.

The relation R is reflexive, symmetric, and transitive so it is an equivalence relation.   □

Note: The preceding example demonstrates that all kinds of properties may be used to define an equivalence relation, and that equivalence classes need not all be the same size.

Example 10.40
Prove that all nilpotent elements of $\mathbb{Z}_n$ are zero divisors.

Proof:

Suppose that $[a] \in \mathbb{Z}_n$ is nilpotent. This means that $a \neq [0]$ and for some finite $k > 1$ we have that $[a^k] = [0]$. Let m be the smallest $m > 1$ such that $[a^m] = [0]$ (by the Well-Ordering Principle such an m exists). Since $m > 1$, a^{m-1} is defined and by definition of m we have that

$[a^{m-1}] \neq [0]$. Thus, since $[0] = [a^m]_n = [a][a^{m-1}]$ we have that $[a]$ multiplied by a non-zero element $[a^{m-1}]$ is zero (mod n) and hence $[a]$ is a zero divisor. □

10.5 PROBLEMS

Problem 10.41
Describe the set of relations on $\mathbb{Z}$ which are both symmetric and anti-symmetric. Hint: this set is infinite and contains one relation with which you are already familiar.

Problem 10.42
Suppose that we call two polygons in the plane *congruent* if one can be translated and rotated to cover exactly the points included in the other. Prove that congruence is an equivalence relation.

Problem 10.43
Prove Proposition 10.13.

Problem 10.44
Count the number of symmetric relation on a three element set.

Problem 10.45
Compute the number of equivalence relations on a three element set.

Problem 10.46
In $\mathbb{Z}_{12}$ find a zero divisor that is not nilpotent.

Problem 10.47
For which natural numbers $n \geq 2$ does $\mathbb{Z}_n$ contain zero divisors?

Problem 10.48
For which natural numbers $n \geq 2$ does $\mathbb{Z}_n$ contain nilpotent elements?

Hint: Consider the prime factorization of n.

Problem 10.49
For each of the following numbers, determine if they are a zero divisor or unit and prove your result. For those that are zero divisors, determine if they are nilpotent. Recall that $[y]_x$ means the equivalence class of y modulo x.

(i) $[12]_{23}$ (ii) $[8]_{26}$ (iii) $[3]_{243}$ (iv) $[7]_{91}$ (v) $[10]_{27}$

Problem 10.50
Suppose that p is a prime number. Prove that Z_p contains no zero divisors.

Problem 10.51
Prove that the set of units in $\mathbb{Z}_n$ is closed under multiplication. That is show that for any two units a, b in $\mathbb{Z}_n$ that ab is a unit.

Problem 10.52
Prove Proposition 10.33.

Problem 10.53
Demonstrate that divisibility does not partially order the integers by finding a violation of one of the three defining properties of a partial order.

Problem 10.54
Prove Proposition 10.37.

Problem 10.55
Draw the Hasse diagram of the divisibility relation for the positive divisors of 96.

Problem 10.56
Draw the Hasse diagram of the divisibility relation for the positive divisors of 30 in the shape of a cube and explain why they form a cube with reference to the prime factorization of 30.

Problem 10.57
Prove that there are no immediately related elements in the partial order $(\mathbb{R}, \leq)$.

Hint: A proof by contradiction is probably easiest.

Problem 10.58
Proposition 10.37 tells us that, given that $(\mathbb{R}, \leq)$ is a partial order so are $(\mathbb{Q}, \leq)$ and $(\mathbb{Z}, \leq)$. Do either of these partial orders have pairs of elements that are immediately related? Prove your answer and, if it is yes, give the pairs of immediately related elements.

Problem 10.59

What maximal subsets of the non-negative integers are totally ordered by divisibility? A set is **maximal** relative to a property if adding any additional element destroys the property.

Hint: These subsets are infinite.

Problem 10.60

Give the Hasse diagram for the sets $S = \{a, b, c\}$ and $T = \{1, 2, 3, 4\}$ under the relation $\subseteq$.

Problem 10.61

The problem assumes you have had a course in calculus. Let $\mathcal{F}$ be the set of differentiable functions defined on the real numbers. Let R be the relation composed of all ordered pairs of the form $(f(x), f'(x))$. Prove or disprove: R is a partial order.

Problem 10.62

Suppose that W is the set of words in the English language, other than proper nouns. Create a relation $R \subset W \times W$ of the form

$$R = \{(s, t) : \text{s appears in t}\}$$

This means that (cat,catastrophe) and (stone,stonewall) are pairs in R.

Prove or disprove:

(i) R is a partial order   (ii) R is a total order.

CHAPTER 11

Number Bases, Number Systems, and Operations

In Section 1.2 we examined the base ten representation of numbers that we use for the real numbers and all the other types of numbers that are subsets of the reals. In this section we are going to take a quick look at the other number bases.

11.1 NUMBER BASES

Definition 11.1 Any whole number may be written as a sum of powers of a positive integer $b > 1$ with coefficients that are natural numbers in the range $0 \leq x < b$. The list of digits in order of descending powers of b, with a decimal point before the digit corresponding to b^{-1} is the **base b representation** of a number. We denote the base of a representation by subscripting the list of digits with b.

Example 11.2
If we do a little calculation we find that

$$87 = 1 \cdot 2^6 + 0 \cdot 2^5 + 1 \cdot 2^4 + 0 \cdot 2^3 +$$
$$1 \cdot 2^2 + 1 \cdot 2^1 + 1 \cdot 2^0$$

so the base 2 representation of 87 is

$$87 = 1,010,111_2$$

The number $\frac{3}{8}$ can be written as

$$\frac{3}{8} = 0 \cdot 2^{-1} + 1 \cdot 2^{-2} + 1 \cdot 2^{-3}$$

and so we write

$$\frac{3}{8} = 0.011_2$$

Let's look at 87 and $\frac{3}{8}$ in base 3.

$$87 = 1 \cdot 3^4 + (0) \cdot 3^3 + (0) \cdot 3^2 + 2 \cdot 3 + (0)1$$

so

$$87 = 10,020_3$$

while

$$\frac{3}{8} = \frac{1}{3} + \frac{1}{27} + \frac{1}{243} + \frac{1}{2187} + \cdots$$

so

$$\frac{3}{8} = 0.1010101\cdots_3 = 0.\overline{10}_3$$

and we see that $\frac{3}{8}$ is a repeating decimal in base 3 even though it is a terminating decimal in base 10: $\frac{3}{8} = 0.375$.

The fact that a fraction is a repeating decimal in one base and a terminating decimal in another leads naturally to the question of which fractions are terminating or repeating in a given base. This question is answered by the following proposition.

Proposition 11.3   Suppose that x and y are integers with $y \neq 0$.
A fraction $\frac{x}{y}$ is a terminating decimal in base b if and only if every prime divisor of y is a prime divisor of b.

Proof:

This proof is left as an exercise.

The base 2 representation is the most famous alternative number base to base 10. It has it's own name: **binary**. It is relatively rare to hear base 2 numbers called "base 2 numbers," numbers expressed in binary are usually called **binary numbers**.

The supply of digits we use in base 10 is adequate for bases $2, 3, \ldots 10$ but as soon as we begin working in base 11 or more the supply of (single character) digits becomes problematic. The only commonly used base above 10 is 16 which is used a good deal in computing. The digits of the base 16 (or **hexidecimal**) system are $\{0, 1, 2, 3, 4, 5, 6, 7, 8, 9, A, B, C, D, E, F\}$. If you look at the source code of a web page you may, for example, see colors specified with hexidecimal numbers.

Example 11.4

$$456 = 1 \cdot 16^2 + 12 \cdot 16^1 + 8 \cdot 16^0$$

so, remembering that 12 is the digit C

$$456 = 1C8_{16}$$

Similarly,

$$10{,}000 = 2 \cdot 16^3 + 7 \cdot 16^2 + 1 \cdot 16^1 + 0 \cdot 16^0$$

so

$$10{,}000 = 2{,}710_{16}$$

Most problems are easier to solve in base 10 because we are used to it. There are occasional problems that get a good deal easier in an alternate number base. For instance, a large number of computer science related problems are significantly easier in binary. In general, however, the mathematics in one number base is equivalent to mathematics in another. We will verify this assertion in some specifics in the problems.

11.2 OPERATIONS AND NUMBER SYSTEMS

We have been using a number of operations, such as addition and multiplication, already. We also have the Boolean operations AND, OR, NOT, IMPLIES, and XOR that were introduced at the beginning of the book. It is assumed that students are at least somewhat familiar with these operations and their properties. In this section we formally define the idea of operation and the associated idea of closure.

Definition 11.5 An n-tuple over a set S is an ordered list of elements from S,

$$(s_1, s_2, \ldots, s_n)$$

The elements of an n-tuple need not be distinct. Ordered pairs are 2-tuples. The set of all n-tuples on a set S is the n-fold Cartesian product $S \times S \times \cdots \times S$ of the set S with itself. This n-fold Cartesian product can also be written S^n.

Example 11.6
The set $\{0, 1\}^3$ is:

$$\{(0, 0, 0), (0, 0, 1), (0, 1, 0), (0, 1, 1),$$

$$(1, 0, 0), (1, 0, 1), (1, 1, 0), (1, 1, 1)\}$$

Definition 11.7   An **n-ary operation** $\circ$ is a function

$$\circ : S^n \to S$$

Operations are sometimes written in a different manner from other functions. If $\circ$ is a 2-ary (or **binary**) operation then instead of the usual functional notation $\circ(x, y)$ we write $x \circ y$. Both multiplication and addition are written in this fashion, which is called **infix notation**. Unary operations, such as negation, are typically written as prefixes: $\neg x$ is the negation of x. Trinary or higher operations are often written using functional notation.

Example 11.8
It is possible to re-write standard arithmetic in prefix notation. For example, the infix expression

$$a * b + 2 * x * y$$

becomes the prefix expression

$$+(*(a, b), *(2, *(x, y)))$$

which is fairly cumbersome. The economy of notation of infix notation is one of its substantial advantages.

Definition 11.9   Suppose that $T \subseteq S$ and that $f : S^n \to S$ is an n-ary operation on S. We say that f is **closed** on T if for each n-tuple $x \in T^n$ we have that $f(x) \in T$.

If we have an operation on a set then it is *by definition* closed on that set. Closure only becomes interesting when we ask if an operation is closed on a subset of the domain of definition of the original operation. For instance, addition is closed on the even integers, since the sum of two even integers is an even integer, but it is not closed on the odd integers.

We now turn to the standard hierarchy of number systems both because they are useful to know and because they form a playground in which to discuss operations and closure. We are already familiar with the natural numbers $\mathbb{N} = \{0, 1, 2, \ldots\}$ and the integers $\mathbb{Z} = \{0, \pm 1, \pm 2, \pm 3, \ldots, \}$.

Definition 11.10   The **rational numbers**, denoted by $\mathbb{Q}$ are the set of ratios of integers. If a and b are integers with $b \neq 0$ and a and b relatively prime then $q = \frac{a}{b}$ is a rational number.

The above defines the *set* of rational numbers and is not the only way to define them. The requirement that a and b be relatively prime forces the rational numbers to be *reduced* fractions, e.g., $\frac{1}{2}$ but not $\frac{2}{4}$ or $\frac{3}{6}$. Another way to define the rationals would be to start with all ratios, reduced or not. Next, define an appropriate equivalence relation, and then declare the rationals to be equivalence classes containing all ratios that have the same reduced form. This method is more cumbersome but also more formal. In the current form we are required to simplify rationals after performing arithmetic operations on them. Addition and multiplication of rationals should already be familiar, but we include them for the sake of completeness.

$$\frac{a}{b} + \frac{c}{d} = \frac{ad + bc}{bd}$$
$$\frac{a}{b} \cdot \frac{c}{d} = \frac{ac}{bd}$$

Notice that the rationals $\frac{n}{1}$ behave exactly as the integers do under addition and multiplication. We say that the rationals *contain a copy of the integers* and, for convenience, we confuse n with $\frac{n}{1}$ so that we may write as integers the copy of the integers inside the rationals.

The next step in the hierarchy of number systems we are building is the **real numbers**, denoted by the symbol $\mathbb{R}$. A formal course in real analysis will really examine the real numbers in depth. Here we will give two informal definitions. First of all, the real numbers are the set of numbers that *can be distances* (including positive, negative and zero distances). A slightly more formal definition is the following, drawing on your calculus training for the notion of limits.

Definition 11.11 Let S be the set of all convergent sequences with members in $\mathbb{Q}$. Then the set $\mathbb{R}$ of **real numbers** is the set of all limits of these sequences.

In other words, the real numbers are the set of all numbers that can be the limit of a convergent sequence of rational numbers.

Proposition 11.12 The rationals are closed, under both addition and multiplication, as a subset of the real numbers.

Proof:

Let $\frac{a}{b}$ and $\frac{c}{d}$ be arbitrary rational numbers. By definition, $\frac{a}{b} + \frac{c}{d} = \frac{ad+bc}{bd}$ and $\frac{a}{b} \cdot \frac{c}{d} = \frac{ac}{bd}$ thus the sum or product of any two rational numbers contains a sum of products of integers in its numerator (in the case of multiplication its a sum with one term), and a product of integers in its denominator. Hence, its numerator and denominator are integers. The product bd is non-zero if and only if both b and d are non-zero, however by definition b and d are non-zero, thus the sum or product of two rational numbers has an integer for a numerator and a non-zero integer for the denominator. When both are reduced to be relatively prime the result is a rational number. $\square$

In Proposition 1.11 we demonstrated that the rational numbers are precisely those with terminating of repeating decimal expansions. We now give a name to the real numbers that are not rationals.

Definition 11.13   The **irrational numbers** are the set $\mathbb{R} - \mathbb{Q}$. That is the set complement of the rationals within the reals.

It is a direct consequence of Proposition 1.11 that the irrational numbers are those with non-repeating decimal expansions.

We already know that there is no real number x so that $x^2 = -1$. When an object does not exist in some context, it is a common algebraic technique to simply declare that the object does exist, give it a name or symbol, and then investigate the consequences. The *complex numbers* are constructed in this manner by creating the symbol i and declaring that $i^2 = -1$. If we take the smallest set containing $\mathbb{R}$ and i that is closed under addition and multiplication we obtain the complex numbers $\mathbb{C}$.

Definition 11.14   The set $\mathbb{C}$ of **complex numbers** are all numbers of the form $z = x + iy$ where $x, y \in \mathbb{R}$ and $i^2 = -1$.

Addition, multiplication, and division of complex numbers work as follows:

$$
\begin{aligned}
(a + bi) + (c + di) &= (a + c) + (b + d)i \\
(a + bi)(c + di) &= (ac - bd) + (ad + bc)i \\
\frac{(a + bi)}{(c + di)} &= \frac{ac + bd}{c^2 + d^2} + \frac{bc - ad}{c^2 + d^2}i
\end{aligned}
$$

Notice that division requires that $c + di \neq 0$.

Outside of algebra the question "is a set closed under an operation" is usually asked of subsets of the complex numbers. Notice that:

$$\mathbb{N} \subset \mathbb{Z} \subset \mathbb{Q} \subset \mathbb{R} \subset \mathbb{C}$$

Each of these sets is closed under addition and multiplication.

Example 11.15
Suppose that m such that $m^2 = 2$ is a complex number and let

$$W = \{q + rm : q, r \in \mathbb{Q}\}$$

Prove that W is closed under addition and multiplication.

Proof:

Let $x = q + rm$ and $y = q' + r'm$ be elements of W. If we add them we obtain $x + y = (q + rm) + (q' + r'm) = (q + q') + (r + r')m$ the last of which is an element of W because $\mathbb{Q}$ is closed under addition. Multiplying x and y we obtain:

$$
\begin{aligned}
xy &= (q + rm)(q' + r'm) \\
&= (qq' + qr'm + rmq' + rmr'm) \\
&= (qq' + rr'm^2) + (qr' + rq')m \\
&= (qq' + 2rr') + (qr' + rq')m
\end{aligned}
$$

We see that the last term is a member of W because $\mathbb{Q}$ is closed under addition and multiplication. Note that in simplifying we exploit the fact that $m^2 = 2$. We thus have that W is closed under addition and multiplication. $\qquad\square$

The set W defined in Example 11.15 is actually a number system. It is derived from the rationals by throwing in the square root of two and then taking all possible products and sums.

The five main number systems defined in the section are all infinite sets. We now prove some classical theorems that throw some light on the relations between these systems. The next theorem is attributed to Pythagoras. History records that this result so impressed him that he sacrificed 100 bulls to the gods in thanks for the insight (Gonick [1980]).

Over the non-negative real numbers we define $\sqrt{x}$ to be a non-negative number y with the property that $y^2 = x$.

Proposition 11.16 $\sqrt{2} \notin \mathbb{Q}$

Proof:

Suppose, by way of contradiction, that $\sqrt{2} \in \mathbb{Q}$. Then there exist $a, b \in \mathbb{Z}$, relatively prime, so that

$$
\begin{aligned}
\sqrt{2} &= \frac{a}{b} \\
b\sqrt{2} &= a \\
(b\sqrt{2})(b\sqrt{2}) &= aa \\
2b^2 &= a^2
\end{aligned}
$$

Since $2b^2$ is even it follows that a^2 is even and so a itself must be even. Thus, there exists $k \in \mathbb{Z}$ such that $a = 2k$. So, $2b^2 = (2k)^2$ and hence $b^2 = 2k^2$. This means that b itself must be even, and so 2 divides both a and b however a and b are relatively prime. This is a contradiction. Therefore, the initial hypothesis that $\sqrt{2} \in \mathbb{Q}$ must be false. $\square$

Another way to state Proposition 11.16 is to say that $\sqrt{2}$ is irrational.

11.3   ADDITIONAL EXAMPLES

Example 11.17
Convert the following base 10 numbers into the specified base:

1. 45 into base 4

2. 66 into base 9

3. 503 into base 2

Solution:

1. 45 : Notice that $4^2 = 16$, and note that $3 \cdot 16 = 48$ which is just slightly too large. So start with $2 \cdot 4^2 = 32$, take 45 and subtract 32 yielding 13. Notice $3 \cdot 4 = 12$ which is just slightly below 13, so take 13 and subtract 12 yielding 1.

   Thus, $45 = 2 \cdot (4^2) + 3 \cdot (4) + 1 = 231_4$.

2. 66 : Notice $9 * 9 = 81$ which is already too large. So start with $7 \cdot 9 = 63$, take 66 and subtract 63 yielding 3.

   Thus, $66 = 7(9) + 3(1) = 71_9$.

3. 503 : Notice $2^9 = 512$ while $2^8 = 256$ so start with 2^8. Take 503 and subtract 2^8 yielding 247, subtract 2^7 from that to yield 119. Take 119 and subtract 2^6 yielding 55, subtract 2^5 yielding 23. subtract 2^4 yielding 7, subtract 2^2 yielding 3, and finally subtract 2 yielding 1.

   Thus, $503 = 2^8 + 2^7 + 2^6 + 2^5 + 2^4 + 0 \cdot 2^3 + 2^2 + 2^1 + 2^0 = 111110111_2$

Example 11.18
Determine whether the negative integers are closed under multiplication and addition.

Solution:

Let A be the set of negative integers. A is not closed under multiplication, a single counterexample is sufficient to demonstrate this. Note that $(-1) \cdot (-2) = 2$, $-1, -2 \in A$ but $2 \notin A$, thus A isn't closed under addition.

The sum of any two negative integers is a negative integer, thus A is in fact closed under addition. To see this note that if $a, b \in A$ then $a < 0$ and $b < 0$ therefore $a + b < 0 + 0 \Rightarrow a + b < 0$.

Example 11.19
Determine which of $\neg$, $\wedge$, $\vee$, and $\rightarrow$ are closed on $\{T\}$, and which are closed on $\{F\}$.

Solution:

By definition, $T \wedge T = T$, $T \vee T = T$, and $T \rightarrow T = T$; thus conjunction, disjunction, and implication are all closed on $\{T\}$. Likewise, $F \wedge F = F$ and $F \vee F = F$ so conjunction and disjunction are closed on $\{F\}$. However, $F \rightarrow F = T$ thus $\rightarrow$ is not closed on $\{F\}$. For negation $\neg T = F$ and $\neg F = T$, thus neither $\{T\}$ nor $\{F\}$ are closed under $\neg$.

11.4   PROBLEMS

Problem 11.20
Convert the following base 10 numbers into the specified base.

1. 27 into base 2

2. 127 into base 3

3. 227 into base 5

4. 5555 into base 16

Problem 11.21
Convert each of these numbers in an alternate base back into base 10.

1. 10101010_2

2. 20122_3

3. $CF5A_{16}$

4. 5545_6

Problem 11.22

Convert the following fractions to the specified base.

1. $\frac{4}{9}$ in base 3.

2. $\frac{1}{4}$ in base 5.

3. $\frac{1}{2}$ in base 3.

4. $\frac{1}{5}$ in base 2.

Problem 11.23

Convert the following numbers to base 10.

1. 0.11011_2

2. $0.\overline{12}_3$

3. $1.\overline{1A}_{16}$

4. $0.\overline{4}_5$

Problem 11.24

If $x = 0.0110111011_2$ then what, in base 2, is $\frac{x}{2}$?

Problem 11.25

Prove Proposition 11.3.

Problem 11.26

Support or refute: constants are zero-ary operations.

Problem 11.27

Let P be the set of all ordered pairs of integers with the second integer non-zero. Define, without reference to reduced fractions or equality to a rational number, an equivalence relation such that the equivalence classes have members that take on a single value when the fraction $\frac{a}{b}$ is reduced where a is the first integer in the ordered pair and b is the second integer in the ordered pair.

Problem 11.28

Define a subset U of the real numbers as follows:

$$U = \{a + b\sqrt{3} : a, b \in \mathbb{Q}\}$$

Show that this set is closed under addition, multiplication, and division other than division by zero.

Problem 11.29
Place the following expressions in infix notation into functional notation:

1. $A + B + C + D + E + F$

2. $2 * (A + B) * (C + D)$

3. 2^{2^n}

4. $\frac{2^{x+y}}{3^{x-y}}$

Problem 11.30
Place the following expressions in infix notation into functional notation:

1. $x^2 + 4x + 6$

2. $(A \wedge B \rightarrow A)$

3. $\sum_{k=1}^{5}(2k)$

4. $(S \cap T)^c \cup (S \cap V)$

Problem 11.31
Show that W from Example 11.15 and U from Problem 11.28 are different as sets and find their intersection. Determine if this intersection is closed under addition and multiplication.

Problem 11.32
Define a subset V of the real numbers as follows,

$$U = \{a + b\sqrt[3]{2} + c\sqrt[3]{4} : a, b, c \in \mathbb{Q}\}$$

Show that this set is closed under addition and multiplication.

Problem 11.33
The **Gaussian integers** are all numbers

$$\{a + bi : a, b \in \mathbb{Z}\}$$

Prove that the Gaussian integers are closed under addition and multiplication.

Problem 11.34

The **norm** or **absolute value** of the complex number $x + iy$ is

$$|x + iy| = \sqrt{x^2 + y^2}$$

A Gaussian integer is **prime** if it cannot be factored into other Gaussian integers with norms that are all larger than 1. Which of $2 + 0i$, $3 + 0i$, $5 + 0i$, $7 + 0i$, and $11 + 0i$ are prime in the Gaussian integers? For each of these that are not prime, provide a factorization.

Problem 11.35

Prove $\sqrt{3} \notin \mathbb{Q}$.

Problem 11.36

Let S be a subset of $\mathbb{R}$. Suppose that $f : S \times S \to S$ is a binary operation which satisfies the following:

- f is an injective function.

- $\{ (f(x, y), x) : x, y \in S \}$ is an equivalence relation.

- $\forall x, y \in S$, $f(x, y)$ is a rational number.

What is the smallest subset S which satisfies these conditions? What is the largest subset S which satisfies these conditions? Explain.

CHAPTER 12

Many Infinities: Cardinal Numbers

It is finally time for us to focus on the main reason set theory was invented in the first place. To discuss and compare infinities. Yes, that's right, infinities plural. This part of set theory is where some of the concepts get very strange. It helps to keep a few things in mind. There are two main uses for numbers: to measure quantity or size and to order things. When we wish to know the number of students in a class we care about that number as a quantity. If we need to systematically respond to requests in terms of priority we care about the order of the requests.

We will examine different infinities by treating them as numbers of a sort, although we will need to update our notion of number to do so. This leads us to two different types of numbers: *cardinal numbers*, to discuss quantity, and *ordinal numbers* to discuss order.

12.1 CARDINALITY

Previously, we defined the cardinality of finite sets by stating that it was the number of elements in a set. Later, we discussed what it means for two (possibly infinite) sets to have the same cardinality, namely that a bijection between those sets exists. It is now time to really properly define the cardinality of sets.

Definition 12.1   Two sets A and B are said to be **similar** or **equinumerous** if and only if there exists a bijection between A and B.

Similar sets have the same cardinality. Now that we have the language to express it, we can see that the notion of similarity (of sets) is in fact an equivalence relation on sets. The identity function which maps every element to itself shows that this relation is reflexive. Since bijections are reversible it's easy to see that the relation is symmetric, and since the composition of bijections is a bijection we see that the relation is transitive.

Definition 12.2   The **cardinality** of a set A, denoted $|A|$, is the equivalence class of A under the similarity relation (for sets).

Definition 12.3   For two sets A and B, A is said to be **dominated by** B, or B **dominates** A whenever it is the case that A is similar to a subset of B. We write $A \leq B$.

While it may not seem it at first this definition could actually be rephrased in the following more intuitive form.

Proposition 12.4   For any two sets A and B, A is dominated by B if and only if there exists an injective function from A to B.

Proof:

Let A and B be sets. Suppose A is dominated by B, then there exists a set $C \subseteq B$ such that A is similar to C, thus there exists a function $f : A \to C$ such that f is a bijection. Since $C \subseteq B$ we may define $g : A \to B$ as the function $g(x) = f(x)$ for all $x \in A$. Note that since f is a bijection it is an injective function and hence g is an injective function.

Now, for the second part of the conditional, suppose that $f : A \to B$ is an injective function. Then $f(A) = \{f(x) \in B : x \in A\}$ is a subset of B. Moreover, $g : A \to f(A)$ given by $g(x) = f(x)$ for all $x \in A$ is an injective function. Noting that for every $y \in f(A)$ there exists an $a \in A$ such that $f(a) = y$ we can conclude that for every $y \in f(A)$ there exists an $a \in A$ such that $g(a) = y$. Therefore, g is both an injection and a surjection and hence a bijection. It follows that A and $f(A)$ are similar sets.   □

The notion of dominance of sets creates a relation on the power set of the universal set.

Definition 12.5   Let $\mathcal{U}$ be the universal set, and let A and B be subsets of $\mathcal{U}$. Then the relation $(\mathcal{P}(\mathcal{U}), \leq)$ where $A \leq B$ if and only if there exists an injection $f : A \to B$, is a partial order that agrees with the notion of less than or equal to for cardinality. Thus, we say that $|A| \leq |B|$ if and only if there exists an injection $f : A \to B$.

Notice that this definition agrees with our previous notion of $|A| \leq |B|$ when the sets are both finite. However, our previous definition for an inequality on the size of sets was only really defined for finite sets. The above is a nice example of a *generalization* where we define or prove something to cover a wider variety of cases than were initially covered.

It is also worth noting that we will use the notation $|A| < |B|$ to refer to the case when A is dominated by B but A and B are not similar. In other words, $|A| < |B|$ refers to the notion of *less than but not equal to*.

12.2   CARDINAL NUMBERS

While we've already been dealing with the natural numbers, by simply refering to them as the counting numbers, or the numbers 0, 1, 2, 3, ..., we do not exactly have a concrete definition for them. In fact, set theory does allow us to do so with cardinal numbers.

We define 0 as the cardinal number of the empty set. Then for any set A we define $|A| + 1$ as the cardinality of the set $A \cup \{c\}$ where $c \notin A$. If we allow for the abuse of notion when $|A| = n$ is a natural number, and we assign to $n + 1$ the appropriate symbol (such as defining 3 as $2 + 1$) we can construct the natural numbers as **cardinal numbers**.

Proposition 12.6   For each natural number n, the cardinal n is the cardinal of the set of natural numbers preceding n in the standard ordering.

Proof by Induction:

The base case is when n is 0. 0 is defined as the cardinal of the empty set. Since no natural numbers precede 0 in the standard ordering it is the case that set of natural numbers preceding 0 is empty. Thus, the base case holds.

Assume, by inductive hypothesis, that the cardinal n is the cardinal of the set $S = \{0, 1, \ldots, n - 1\}$. Notice that $n \notin S$ and so by the definition of $|S| \cup \{n\}$ we have that

$$n + 1 = S \cup \{n\} = \{0, 1, \ldots, n - 1\} \cup \{n\} = \{0, 1, \ldots, n - 1, n\}$$

It follows by induction that the cardinal n is the cardinal of the set of natural numbers preceding n in the standard ordering.   □

The natural numbers are known as the **finite cardinals**. In fact, when we refer to *finite sets* we are referring to a set which has a finite cardinal as its cardinal.

The standard definition of less than or equal to ($\leq$) on natural numbers corresponds to the $\leq$ symbol in the dominance of sets relation on sets. So that $\leq$ on cardinal numbers is the familiar less than or equal to on natural numbers when the cardinals are natural numbers. However, we need an additional result to prove this.

Proposition 12.7   For any set A and natural number n, if $|A| = n$ then A is not similar to a proper subset of itself.

Proof by Induction:

The base case is when n is 0. In this case the only option for A if $|A| = n$ is that A is the empty set. Since the empty set cannot have a proper subset the result holds for $n = 0$.

Assume by inductive hypothesis that if $|A| = n$ then A is not similar to a proper subset of itself.

Inductive Step: Let us assume, by way of contradiction, that B is a set with cardinality $n + 1$ and that there exists C a proper subset of B such that B is similar to C. WOLOG $B = A \cup \{b\}$

where $|A| = n$ and $b \notin A$. Since B and C are similar there exists a bijection $f : B \rightarrow C$.

There are two options either $f(b) = b$ or $f(b) \neq b$. In either case we may replace f with a function $h : B \rightarrow C$ which is identical to f except that if $f(b) \neq b$ then $h(b) = b$ and $h(f^{-1}(b)) = f(b)$. h is still a bijection of B and C.

Consider $g : A \rightarrow C - \{b\}$ given by the restriction of h to A rather than B (in other words g is identical to h except g isn't defined for the value b). The function g must still be a bijection since removing a single value does not change this.

However, $C - \{b\}$ is a proper subset of A since C is a proper subset of B and $A = B \cup \{b\}$ but by the inductive hypothesis A is not similar to a proper subset of itself. This is a contradiction, thus the result holds for $n + 1$.

The proof by contradiction used in the inductive step combined with the result of the base case implies that the proposition follows by induction. □

Notice that in the preceding proof we actually had to make use of a combination of a proof by induction and a proof by contradiction. This is a valid proof technique but the resulting proof may at times be difficult to follow. It helps to point out the individual steps, like the inductive hypothesis and when we are using a proof by contradiction.

Proposition 12.8   The cardinal ordering and the natural number ordering agree on $\mathbb{N}$. That is, for $x, y \in \mathbb{N}$, $x \leq y$ if and only if for sets A and B with $|A| = x$ and $|B| = y$ there is an injective function from A to B.

Proof:

This proof is left as an exercise.

Hint: You will require the results of Propositions 12.6 and 12.7.

12.3   TRANSFINITE CARDINALS

We have established a general notion of cardinal numbers and looked at the case of finite cardinals. Now, is the time to turn our attention to *transfinite* cardinals. In fact, we can now give a proper definition of an infinite set.

Definition 12.9   If the cardinal of a set X is not a finite cardinal, then X is said to be an **infinite set** and X is said to have a **transfinite** or **infinite** cardinal number. In other words, if a set X does not have a natural number as its cardinal number, then it has an infinite cardinal number.

The set $\mathbb{Q}^+$ is the set of positive rational numbers. In general, superscripting a number system that has positive and negative numbers with a plus denotes the subset of positive members. Superscripting the natural numbers with a plus symbol indicates the set $\mathbb{N} - \{0\}$. The next theorem is an example of a counter-intuitive result: there are the same number of natural numbers and positive rationals.

Theorem 12.10 $|\mathbb{N}^+| = |\mathbb{Q}^+|$

Proof:

Consider the following infinite array, with rows and columns indexed by the positive natural numbers:

	1	2	3	4	$\cdots$
1	1/1	1/2	1/3	1/4	$\cdots$
2	2/1	2/2	2/3	2/4	$\cdots$
3	3/1	3/2	3/3	3/4	$\cdots$
4	4/1	4/2	4/3	4/4	$\cdots$
$\cdots$	$\cdots$	$\cdots$	$\cdots$	$\cdots$	$\cdots$

The cells of the array contains the ratio a/b where a is the row index and b is the column index. Reduce all the ratios so that the cells of the array contain a rational number. Notice that every positive rational number appears an infinite number of times in this array. Now we number the cells of the array by sweeping back and forth on diagonals as follows:

	1	2	3	4	$\cdots$
1	1	2	6	7	$\cdots$
2	3	5	8	$\cdots$	$\cdots$
3	4	9	$\cdots$	$\cdots$	$\cdots$
4	10	$\cdots$	$\cdots$	$\cdots$	$\cdots$
$\cdots$	$\cdots$	$\cdots$	$\cdots$	$\cdots$	$\cdots$

This is an ordering of the cells of the array called a *dovetail* ordering. Since each of the diagonals is finite, this order assigns a positive natural number to every cell of each diagonal and so creates a bijection of the positive natural numbers with the cells of the array.

Define an equivalence relation on the cells as follows: two cells are related if they contain (different versions of) the same rational number. It is obvious that this is an equivalence relation.

Take the set of natural numbers assigned to a given equivalence class. This set is a non-empty set of positive numbers and so, by the well ordering principle, contain a least member.

Define a function $f : \mathbb{Q}^+ \to \mathbb{N}^+$ that assigns to each positive rational the least natural number assigned to a cell in the equivalence class of cells associated with that rational. Since each natural number is assigned to a unique cell in the array it follows that f is an injection.

By Definition 12.5 the existence of such an injection tells us that $|\mathbb{N}^+| \geq |\mathbb{Q}^+|$. It is easy to find an injection $g : \mathbb{N}^+ \to \mathbb{Q}^+$ by taking $g(n) = n/1$ and so we likewise have $|\mathbb{Q}^+| \geq |\mathbb{N}^+|$. It follows that $|\mathbb{N}^+| = |\mathbb{Q}^+|$. $\square$

Note that in both injections one could assign 0 to 0 and hence the theorem is true for the non-negative rationals and the natural numbers as well.

Between any two natural numbers are an infinite number of rationals and yet the preceding theorem tells us there is a bijection of the positive rational with the natural numbers. In some ways this seems horribly, horribly wrong. The best view one can take of this result is that it demonstrates the awesome power of proof to let us discover facts that initially appear to our intuition as nonsense. There is an obvious Corollary to this theorem.

Corollary 12.11 $|\mathbb{Q}| = |\mathbb{Z}|$

Proof:

Theorem 12.10 tells us that there is a bijection of $\mathbb{Q}^+$ and $\mathbb{N}^+$. Take this bijection and construct a bijection of the negative rationals and the negative integers by negating both coordinates of each ordered pair. Take the union of these two bijections (recall they are sets of ordered pairs) and add in the ordered pair (0,0). The results is a bijection of $\mathbb{Q}$ and $\mathbb{Z}$. $\square$

We see that two infinite sets that fit into the number line in very different ways can be the same size. While it has been some time since we explicitly dealt with the ordered pair definition of a function notice that it is a beautifully simple way of constructing a new function out of other functions. This is an example of why the set theoretic definition of functions, relations, and other mathematical concepts have made their way into modern advanced mathematics. The set theoretic concepts are a useful way of discussing, in precise language, simple or complicated structures which come up again and again.

Definition 12.12   Any infinite set that is the same size as the natural numbers is said to be **countably infinite**. Any set that is finite or countably infinite is said to be **countable**. The symbol $\aleph_0$ (pronounced aleph naught) is used to denote the cardinality of a countably infinite set.

Countable infinities have the smallest cardinality possible for an infinite set. If it were the case that all infinite sets had the same cardinality, there wouldn't really be much of a point to cardinal numbers. That is, however, not the case. There are infinities which are in a sense *larger* than other infinities.

Proposition 12.13   Let $a, b \in \mathbb{R}$ with $a < b$, then (a, b) has the same cardinality as $\mathbb{R}$.

Proof:

All we need to do is construct a bijection between (a, b) and $\mathbb{R}$. Consider the function $f : \mathbb{R} \to (a, b)$ given by $f(x) = \frac{(b-a)x}{2|x|+2} + \frac{(a+b)}{2}$. As $x \to \infty$, $f(x) \to b$, and as $x \to -\infty$, $f(x) \to a$, and for all $x \in \mathbb{R}$, $a < f(x) < b$. Thus, f is surjective. If $f(x) = f(y)$ then $\frac{(b-a)x}{2|x|+2} + \frac{(a+b)}{2} = \frac{(b-a)y}{2|y|+2} + \frac{(a+b)}{2}$

$\Rightarrow \frac{x}{|x|+1} = \frac{y}{|y|+1}$ this means x and y are either both positive or both negative, and $x|y| + x = y|x| + y$, but since x and y have the same sign $x|y| = y|x|$ therefore $x = y$. Thus, f is an injective function. Thus, f is a bijection and (a, b) and $\mathbb{R}$ have the same cardinality.   □

The set of real numbers has the same size as any open interval of the real numbers, we call this cardinal **the continuum**. Sets whose cardinality is at least as large as the real numbers are said to have *the power of the continuum*.

The following theorem (and associated proof) is sufficiently ingenious to warrant its own name.

Theorem 12.14   Cantor's Diagonal Theorem
$\mathbb{R}$ *is not a countable set.*

Proof:

Suppose, by way of contradiction, that $|\mathbb{R}| = |\mathbb{N}|$. Then there exists a bijection $f : \mathbb{R} \to \mathbb{N}$ that assigns a natural number to each real number. Order the real numbers according to the natural numbers assigned to them by f so that we list the set $\mathbb{R}$ in the order $r_1, r_2, r_3, \ldots$. We now use the decimal expansions of the real numbers, as given in Definition 1.4. Construct a real number x as follows. The nth digit of x equals nine minus the nth digit of the decimal expansion of r_n. Insert a decimal before the first digit in the sequence of digits describing x. Clearly, x is a real number and so must appear in the list $r_1, r_2, r_3, \ldots$. By construction the digits of x differ, in at least one position, from the digits of every real number on the list. This means that x is not in the list of reals. We thus achieve a contradiction from which we conclude that our hypothesis, $|\mathbb{R}| = |\mathbb{N}|$, which permitted us to construct the list, is false. This implies that no bijection exists between the natural numbers and the real number and so the two sets must not have the same cardinality. It is easy to find an injection of the natural numbers which tells us that $|\mathbb{N}| \leq |\mathbb{R}|$, simply use the identity function, establishing that the difference in cardinality goes in the order $|\mathbb{N}| < |\mathbb{R}|$. $\qquad\square$

Infinite sets that do not have the same cardinality as $\aleph_0$ are said to be *uncountable*. Notice the trouble we went to in Chapter 1 to get the decimal representations of real numbers to be unique (exactly one representation per real number) pays off in the proof of Cantor's diagonal theorem. If there had been more than one representation of some real number then the number x might have been a different representation of some number already on the list rather than a unique number.

With his diagonal theorem Cantor demonstrated that there are very different types of infinities, countable ones and uncountable ones. However, Cantor also showed another striking result which in fact demonstrates that there are an infinite number of distinct infinities.

Theorem 12.15   Cantor's Theorem
For every set A. $|A| < |\mathcal{P}(A)|$. In other words, A is dominated by the power set of A, but the two sets are not similar.

Proof:

The function $h : A \rightarrow \mathcal{P}(A)$ which takes each element $a \in A$ to the single element set $\{a\} \in \mathcal{P}(A)$ (i.e., $f(a) = \{a\}$), is an injective function. This demonstrates that $|A| \leq |\mathcal{P}(A)|$.

We will deal with the case $A = \emptyset$ separately (so that later on we can assume the set A contains an element). The empty set has exactly one subset, itself. However, this means that $|\mathcal{P}(\emptyset)| = 1$ and $|\emptyset| = 0$ so $|\emptyset| < |\mathcal{P}(\emptyset)|$.

In order to show that A and $P(A)$ are not similar we will proceed with a proof by contradiction when $A \neq \emptyset$.

Assume, by way of contradiction, that A and $\mathcal{P}(A)$ are similar. Thus, there exists a bijective function $f : A \rightarrow \mathcal{P}(A)$. Consider the set $B = \{a \in A : a \notin f(a)\}$, the set of all elements in A which f does not map to a set which contains the initial element. Notice that by definition B is a subset of A and hence $B \in \mathcal{P}(A)$. So, since f is a bijective function, there must be some element $c \in A$ such that $f(c) = B$. Either $c \in B$ or $c \notin B$. If $c \in B$ then by definition $c \notin f(c)$, but since $f(c) = B$ this means $c \notin B$. This is a contradiction. If $c \notin B$ then by definition $c \in f(c)$ but since $f(c) = B$ this means $c \in B$. This is also a contradiction. Thus, A and $\mathcal{P}(A)$ cannot be similar and it follows that $|A| < |\mathcal{P}(A)|$. $\qquad\square$

Cantor's Theorem is worth examining in some detail. The proof is not long but it provides a rather deep result. The proof is made up of a direct portion and a proof by contradiction portion. The real genius lies in the proof by contradiction and the somewhat self-referential nature of the set B. It will become apparent only when we later get to the chapter on axiomatic set theory why it matters that we constructed B explicitly as a subset of the set A. The real beauty of Cantor's Theorem though is in it's simplicity.

Corollary 12.16   *There are an infinite number of different infinite cardinals.*

Proof:

Define the indexed successor of a set S by the power set of S. Index the sequence of indexed successors, by beginning with the set of natural numbers, $\mathbb{N}$ with index 0 and continuing the index in the natural numbers, so that 1 indexes the power set of $\mathbb{N}$ and so on. By Cantor's Theorem the indexed successor of each set is *larger* than the set which preceded it. Noting that $\mathbb{N}$ is an infinite set to begin with we have that the index sequence indexes a sequence of infinite sets whose cardinalities all differ. $\qquad\square$

12.4   CARDINAL RESULTS

Before stating some of the more interesting results about cardinal numbers, we need to specify another axiom (or principle) which we take to be true.

Definition 12.17   **Axiom of Choice** For any set X $\exists$ a function $f : (\mathcal{P}(X) - \emptyset) \to X$ s.t. $\forall A \in (\mathcal{P}(X) - \emptyset)$, $f(A) \in A$. Such a function is called a **choice function**.

The Axiom of Choice will be discussed in greater detail in Chapter 14, but for now the important thing to note is that it gives us a way of taking a collection of subsets of some set and picking a element from each subset to act as a representative of the subset.

Proposition 12.18   Every infinite set contains a subset with cardinality $\aleph_0$.

Proof:

Let S be an infinite set and let $f : (\mathcal{P}(S) - \emptyset) \to S$ be a choice function for S. Let $g : \mathbb{N} \to S$ be the function defined as follows $g(0) = f(S)$, and for $n \geq 0$, $g(n + 1) = f(S - \{g(0), \ldots, g(n)\})$. Note that g is well defined: $g(0)$ is explicitly defined and $g(n + 1)$ is defined in terms of $g(m)$ where $m \leq n$. Since S is an infinite set $S - \{g(0), \ldots, g(n)\}$ is never the empty set for any n. Note that $g(S)$ is a subset of S. Now we will demonstrate that $h : \mathbb{N} \to g(S)$ given by $h(x) = g(x)$ for all $x \in \mathbb{N}$ is in fact a bijection.

By definition, h is surjective, since we defined the range as the image of g and both h and g evaluate to the exact same elements. Suppose that for some $m, k \in \mathbb{N}$, $h(m) = h(k)$ $\Rightarrow g(m) = g(k)$. WOLOG assume that $m \geq k$ then $g(m) \in f(S - \{g(0), \ldots, g(m-1)\}$ which means that $g(m)$ is not equal to $g(k)$ for all $k \leq m - 1$. Thus, $g(m) = g(k) \Rightarrow m = k$. Therefore, g is an injective function.

Since g is a surjective and injective function it is a bijection and thus S is similar to a subset with cardinality $\aleph_0$. □

If X is a countably infinite set (with cardinality $\aleph_0$) then $\aleph$ is the cardinality of $\mathcal{P}(X)$. To abuse the notation slightly $\aleph = |\mathcal{P}(\aleph_0)|$. Cantor's Theorem tells us that $\aleph$ is a larger cardinal than $\aleph_0$. It turns out that $\aleph$ is a cardinal we are already familiar with. Before we get into that, there is a quick definition we need.

Definition 12.19   Let X be a set. A **characteristic function** is a function $f : X \to \{0, 1\}$. The values $x \in X$ for which $f(x) = 1$ define the **characteristic set** of the function. Characteristic functions are sometimes called *indicator functions*.

Proposition 12.20   The cardinality of the set of real numbers is $\aleph$.

Sketch of Proof:

The proof has three parts, we will provide a sketch of a proof, and leave the filling in of the details as an exercise. First, we need to show that for a set X, $\mathcal{P}(X)$ has the same cardinality as the set of all *characteristic functions*. There is an easy way to do that. Second, we demonstrate that $|\mathbb{R}| \leq \aleph$. We do this by looking at the function $g : \mathbb{R} \to \mathcal{P}(\mathbb{Q})$ given by $g(x) = \{y \in \mathbb{Q} : y < x\}$ and showing that it's injective. Third, the hardest part, is showing that $\mathcal{P}(\aleph_0) \leq \aleph$.

We will use the fact that the power set of a set has the same cardinality as the set of all characteristic functions of the set. The key is to construct an injection from the set of all characteristic functions on the positive integers to the open interval $(0, 1)$. To do this we use the function h from the set of all characteristic functions on $\mathbb{Z}^+$ to $(0, 1)$ given by the following.

For every characteristic function $f : \mathbb{Z}^+ \to \{0, 1\}$, $h(f) = 0.f(1)f(2)f(3)\ldots$. It's easy to verify this is an injection. □

12.5   THE CANTOR SET

The Cantor set is a strange sort of set discovered by Henry John Stephen Smith in 1874, but introduced to the general public and elaborated upon by Georg Cantor whose name ended up being attached to it. The set exhibits properties which seems as though they should be impossible. Its existence indicates that an intuitive understanding of which sets have cardinality $\aleph_0$ and which have cardinality $\aleph$ may be misleading.

There are multiple equivalent ways to define this strange set which is simply called: **The Cantor Set**. We will use the iterative definition as it highlights the bizarre nature of the set.

Definition 12.21   The **Cantor set** is defined iteratively. Start with the interval $[0,1]$, call it C_0. Removed the open interval $\left(\frac{1}{3}, \frac{2}{3}\right)$ that comprises the middle third of this interval. This leaves two intervals which make up the set C_1. Remove the open middle-third of both the intervals in C_1 in the same fashion to get C_2, which contains four intervals. Continue in this fashion, creating C_3, C_4, etc. The Cantor set C is the intersection of all C_n, $n \in \mathbb{N}$.

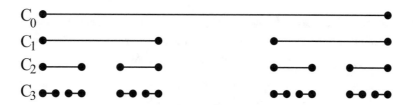

It's worth noting that $\frac{1}{3} \in C$ as is $\frac{2}{3}$ and any other endpoint of the closed intervals of C_k for any $k \in \mathbb{N}$. So it's clear that C does contain at least a countable number of elements.

It is not too hard to see that the length of C_0 is 1 and the total length of C_{i+1} is two-thirds the total length of C_i. The total length L of the set C is at most that of any of the sets that are intersected to create it which tells us that $L \leq (2/3)^n$ for all $n \in \mathbb{N}$.

Taking a quick limit we deduce that $L = 0$ and so the total length of the Cantor set is zero. What, however, is its cardinality?

At each step, we remove a middle third from all present intervals and so create two new intervals. A point in the Cantor set may thus be specified as a choice, in each C_n, of going to the left or right remaining piece from the interval chosen in, C_{i-1}. These choices yield a bijection of the Cantor set with the set of functions $f : \mathbb{N} \to \{0, 1\}$, in other words, with countably infinite sequences of zeros and ones.

There is an obvious surjection from these sequences to the set of all binary decimal numbers in the range $0 \leq x \leq 1$ (put a decimal point in front of the sequence of 0s and 1s). We know that the interval $[0, 1]$ is an uncountable set. We thus see that C is an uncountably infinite set since it's cardinality must be at least as large as the cardinality of $[0, 1]$.

So C has cardinality $\aleph$ but a total length zero. This seems as though it shouldn't be possible. There are no open intervals anywhere in the set, yet it has an uncountable cardinality. The Cantor set demonstrates that intuition about which cardinalities are assigned to subsets of the real numbers is possibly misleading.

12.6   ADDITIONAL EXAMPLES

Example 12.22
Prove that the union of a countable number of countable sets is, itself, countable.

Proof:

We need to show that the union of a countable collection of sets, each of which is a countable set, is itself countable. To do this we will use a dovetail ordering. We begin by constructing a matrix, and sweeping along the rows and columns, like below.

	1	2	3	4	⋯
1	1	2	6	7	⋯
2	3	5	8	⋯	⋯
3	4	9	⋯	⋯	⋯
4	10	⋯	⋯	⋯	⋯
⋯	⋯	⋯	⋯	⋯	⋯

This demonstrates that it is possible to index two countably infinite sequences, the rows and columns, using a single countably infinite sequence. We call such an ordering of the natural numbers a dovetail ordering.

By noting that we can use one of the two infinite sequences indexed by the dovetail ordering to indicate which of the countable sets we wish to index and using the other of the two infinite sequences to indicate which element in the indexed set to consider we can index every element in all of the countably infinite collection of sets.

Now, it is possible that some of the sets are finite, or that we are taking the union of a finite number of sets, but in this case we will simply assign a dummy element (or dummy set) to these indexed set and element pairs. The union itself is at most as large (in terms of cardinality) as the enumeration with possible dummy elements.

Thus, the enumeration of the elements of the countable union of countable sets indicates that the union is countable. $\qquad\square$

Example 12.23
Let C be the Cantor set. Show that if $x \in C$ then $\frac{x}{3} \in C$.

Solution:

Note that when we remove the open interval in C_1 then the interval $[0, \frac{1}{3}]$ is simply a scaled, similar version of C_0 but with the index off by one. Similarly, when we remove the open intervals from C_2, C_3 etc. the interval $[0, \frac{1}{3}]$ remains a scaled, similar version of the entire interval $[0, 1]$ but one step behind in terms of removing the open intervals.

If $x \in C$ then x is never removed at any step along the way, since the portion of the C_k in the interval $[0, \frac{1}{3}]$ is simply a scaled version of the interval $[0, 1]$ for C_{k-1} this means that $(\frac{1}{3})x$ is also not removed at any step along the way. Thus, if $x \in C$ then $\frac{x}{3} \in C$.

12.7 PROBLEMS

Problem 12.24
Prove that the set V from Problem 11.32 is countable.

Problem 12.25
Prove that $|\mathbb{N}| = |\mathbb{Z}|$.

Problem 12.26
Prove that $|\mathbb{N}| = |\mathbb{Q}|$.

Problem 12.27
Suppose that S is a countable set. Prove that S^n is also countable.

Problem 12.28
Let C be the Cantor set. Show that if $x \in C$ then $(1 - x) \in C$.

Problem 12.29
Show that there are only countably many points in the Cartesian plane that have both of their coordinates rational. These points are called *rational points*.

Problem 12.30

Suppose that the universal set is $\mathbb{R}$. If $r \in \mathbb{Q}$ and $s \in \mathbb{Q}^c$ prove that $r + s \in \mathbb{Q}^c$.

Problem 12.31

Using the iterative definition of the Cantor set, $\mathcal{C}$, show that an alternate way to define $\mathcal{C}$ would be to say that:

$\mathcal{C} = \{x \in [0, 1] : x \text{ has a ternary expansion which does not require 1s }\}.$

In this definition we assume that the ternary expansions are not unique. i.e., $\frac{1}{3} \in \mathcal{C}$ even though $\frac{1}{3} = 0.1_3$ because $\frac{1}{3} = 0.22222\ldots_3$ as well.

Problem 12.32

Let L be the set of lines in the Cartesian plane that contain no rational points. Answer the following questions, proving your answer.

   (i) Is the cardinality of L finite, countably infinite, or uncountably infinite?

   (ii) Does L contain a line with slope 1?

   (iii) Let $r \in \mathbb{Q}$, does L contain a line with slope r?

Problem 12.33

Prove that no equilateral triangle with side length one in the Cartesian plane has vertices that are all rational points.

Problem 12.34

Read Problem 10.58. What condition on a subset of $S \subseteq \mathbb{R}$ forces each element of $(S, \leq)$ to be immediately comparable to at least one other element? Prove your answer.

Problem 12.35

Since the rational numbers are countable, we can find a bijection of $f : \mathbb{N} \to \mathbb{Q}$ that permits us to list the rational numbers in the order $f(0), f(1), f(2), \ldots$. Let I_n be the interval of real numbers

$$[f(n) - 2^{-n}, f(n) + 2^{-n}]$$

Let

$$S = \cup_{n=0}^{\infty} I_n$$

Prove that there are an infinite number of real numbers outside of S.

Problem 12.36

Let A be the set of real numbers that are roots of polynomials with integer coefficients. Prove that A is countable.

Problem 12.37

The *procedurally representable numbers* are those real numbers r for which there exists, in abstract, a computer program that can type out the first n digits of r in finite time for all natural numbers n. Prove that the procedurally representable numbers are a countable set.

Problem 12.38

Read Problem 12.37. Prove that the procedurally representable numbers are closed under addition and multiplication.

Problem 12.39

Read Problem 12.37. Prove that the members of the set $\mathbb{Q}$ are procedurally representable.

Problem 12.40

Consider the partial order

$$(\mathcal{P}(\mathbb{R}), \subseteq)$$

Is each member immediately related to another?

Problem 12.41

Fill in the details for the proof of Proposition 12.20, using the sketch of the proof provided.

CHAPTER 13

Many Infinities: Ordinal Numbers

We've examined the abstracted notion of the size of a number with cardinal numbers, so now we examine the abstracted notion of order.

13.1 PARTIALLY ORDERED SETS REVISITED

This entire chapter concerns partially ordered sets, so we include a brief reminder and some notes on terminology. A *partially ordered set* (or partial order) is a relation, $(S, \leq)$, which is: *reflexive* ($\forall a \in S$, $a \leq a$), *antisymmetric* ($\forall a, b \in S$, if $a \leq b$ and $b \leq a$, then $a = b$), and *transitive* ($\forall a, b, c \in S$, if $a \leq b$ and $b \leq c$, then $a \leq c$).

We adopt the following notation: $a \leq b$ is read as a precedes b (or b succeeds a). If $a \leq b$ and $a \neq b$, then $a < b$, which is read a strictly precedes b (and b strictly succeeds a).

If $a \leq b$ or $b \leq a$, then we say that a and b are *comparable*. If $(S, \leq)$ is a partially ordered set and for all $a, b \in S$, a and b are comparable, then $(S, \leq)$ is said to be a **totally ordered set** or **total order**. Total orders are sometimes called **linear orders** or **linearly ordered sets**.

13.2 PARTIALLY ORDERED SETS EXPANDED

This section introduces some new terminology and results for partially ordered sets.

Definition 13.1   Let (S, R) be a partially ordered set, with the relation R. Let $A \subseteq S$. Then the **induced order**, (A, R_A), is the partial order $R_A = R \cap (A \times A)$. In other words, if $x, y \in A$, then $x R y$ if and only if $x R_A y$. The induced order (A, R_A) is also called the **restriction of S to A**. The symbol R is often used instead of R_A for the sake of convenience.

Definition 13.2   Let $(S, \leq)$ be a partially ordered set. A subset $A \subseteq S$ is said to be a **chain in S** if the induced order $(A, \leq)$ is a total order.

Notice that in a totally ordered set every subset is a chain. However, in a partial order in general, subsets are not necessarily chains.

What follows are some convenient ways to construct new orders out of existing orders.

Definition 13.3   Let (S, R) be a partial order with relation R. The **dual order** of (S, R) is the partial order (S, R') given as follows: $\forall a, b \in S, (a, b) \in R$ if and only if $(b, a) \in R'$.

Just because we called the dual order a partial order doesn't mean it actually is one, you are asked to prove that the definition actually defines a partial order in Problem 13.36.

Definition 13.4   Let I be a totally ordered index set. Let $\{C_i : i \in I\}$ be a collection of pairwise disjoint totally ordered sets (i.e., $C_i \cap C_j = \emptyset$ when $i \neq j$).

The union $A = \cup_{i \in I} C_i$ along with the following partial order is the **concatenation order** or **summation order**: $x \leq y$, if $x \in C_i$, $y \in C_j$ and $i < j$ in the index set ordering, or $x, y \in C_i$ and $x \leq y$ in the partial order $(C_i, \leq)$.

The concatenation order can be visualized by listing ordered elements of C_i before ordered elements of C_j when $i < j$ and separating the various C_i and C_j by semicolons.

Example 13.5
Let E be the even non-negative integers, let A be the letters in the alphabet. Using the standard order:
$$E \cup A \cup (\mathbb{N} - E) = \{0, 2, 4, 6, 8, 10, \ldots; a, b, c, \ldots y, z; 1, 3, 5, 7, 9, \ldots\}$$

Proposition 13.6   The concatenation order is a total order.

Proof:

First, we need to show that the concatenation order is in fact a partial order.

(Reflexivity) Note that for any $x \in A$ there exists and $i \in I$ such that $x \in C_i$. Since C_i is a total order it follows that (x, x) is in the relation for C_i and hence in the relation for A. Thus, $(A, \leq)$ is reflexive.

(Anti-symmetry) Suppose $a, b \in A$, with $a \leq b$ and $b \leq a$. Since $a \leq b$ we have that either **(i)** $a \in C_i$ and $b \in C_j$ with $i < j$, or **(ii)** $a, b \in C_i$ for some $i \in I$ and $a \leq b$ in the total order $(C_i, \leq)$. In the case of **(i)** $b \leq a$ implies that $j < i$ and $i < j$ which is a contradiction. Thus, only case **(ii)** remains a valid option. In this case, the fact that C_i is itself a partial order indicates that if $a \leq b$ and $b \leq a$ then $a = b$. Thus, $(A, \leq)$ is anti-symmetric.

(Transitivity) Suppose $a, b, c \in A$ and $a \leq b$ and $b \leq c$. There are two cases, either (i) there exists a single index i such that $a, b, c \in C_i$ or (ii) there exist indices i, j such that $i < j$ and $a \in C_i$ and $c \in C_j$. Note that b is not guaranteed to be in either of these sets, but the fact that $a \leq b$ and $b \leq c$ indicates that the indices for a and c must either be the same or $i < j$. For (i) if $a, b, c \in C_i$ for some $i \in I$ then the fact that each $(C_i, \leq)$ is a partial order indicates that the relation must be transitive. For (ii), since $a \in C_i$ and $c \in C_j$ with $i < j$ then by definition $a \leq c$, and thus transitivity holds. Thus, the concatenation order is reflexive, anti-symmetric and transitive, and hence a partial order.

(Comparability) To see that it's a total order simply note that for any $a, b \in A$ there exists $i, j \in I$ (possibly the same) such that $a \in C_i$ and $b \in C_j$. If $i < j$ or $j < i$ then $a \leq b$ or $b \leq a$, respectively, and hence a and b are comparable. If $i = j$ then $a, b \in C_i$ and the fact that C_i is a total order indicates that all elements of C_i are comparable and hence a and b are comparable. Thus, the concatenation order is a total order. $\qquad\square$

13.3   LIMITING ELEMENTS

Some partially ordered sets have limiting elements. These are elements which may be the minimal or maximal, or first or last element.

Definition 13.7   Let $(X, \leq)$ be a partially ordered set. An element $y \in X$ s.t. $y \leq a$ for all $a \in X$ is called a **first element** of X. An element $z \in X$ s.t. $b \leq z$ for all $b \in X$ is called a **last element** of X.

Not every partially ordered set has a first element, nor does every partially ordered set have a last element. The natural numbers (with the usual order) $(\mathbb{N}, \leq)$ has 0 as a first element but has no last element. The rational numbers (with the usual order) $(\mathbb{Q}, \leq)$ have neither a first nor last number. However, if a set has a first element it can only have at most one first element. Similarly a set can have at most one last element. The uniqueness of the first and last elements in demonstrated in the following proposition.

Proposition 13.8   Let X be a partially ordered set.

(i) If there exists $a, b \in X$ s.t. $\forall y \in X$, $y \leq a$ and $y \leq b$, then $a = b$.

(ii) If there exists $c, d \in X$ s.t. $\forall y \in X$, $c \leq y$ and $d \leq y$, then $c = d$.

Proof:

Let X be a partially ordered set.

(i) Suppose there exists $a, b \in X$ s.t. $\forall y \in X$, $y \leq a$ and $y \leq b$. Since $a, b \in X$ we have that $a \leq b$ and $b \leq a$. Since X is a partially ordered set, the relation $\leq$ is anti-symmetric, thus when $a \leq b$ and $b \leq a$ we have that $a = b$.

(ii) Suppose there exists $c, d \in X$ s.t. $\forall y \in X$, $c \leq y$ and $d \leq y$. Since $c, d \in X$ we have that $c \leq d$ and $d \leq c$. Since X is a partially ordered set, the relation $\leq$ is anti-symmetric, thus when $c \leq d$ and $d \leq c$ we have that $c = d$. □

Definition 13.9   Let $(X, \leq)$ be a partially ordered set. An element $y \in X$ is called a **maximal element** if $y \leq a$ implies that $a = y$. An element $z \in X$ is called a **minimal element** if $b \leq z$ implies that $b = z$.

Maximal elements are elements that no other element strictly succeeds (is greater than). Minimal elements are elements which no other element strictly precedes (is less than). Unlike first and last elements, partially ordered set can have multiple maximal or minimal elements.

Example 13.10   Consider the set $S = \{x > 1 : x$ is an integer $\}$. Ordered by the divisibility relation (that is $a \leq b$ if $a|b$).

Every prime number is a minimal element because it is not divisible by a positve integer other than itself and 1 (which we've explicitly excluded). None of the prime numbers are a first element because none of the prime numbers are comparable.

However, the set $S \cup \{1\}$ ordered by divisibility has a unique minimal element (and a first element): 1.

A finite set always has at least one minimal and one maximal element. If a set has a first element that element must be a minimal element. If a set has a last element that element must be a maximal element.

Totally ordered sets behave differently than general partially ordered sets. If a total order has a maximal element that element is a last element. Similarly, if a total order has a minimal element that element is a first element. A finite total order must have a first and last element.

Definition 13.11   Let $(X, \leq)$ be a partially ordered set, and let $Y \subseteq X$. An element $a \in X$ is called an **upper bound** of Y if $y \leq a$ for all $y \in Y$. An element $c \in X$ such that $c \leq a$ for all upper bounds a, is called the **supremum** of Y, and is denoted $sup(Y)$. The supremum of Y is sometimes known as the **least upper bound** of Y.

Definition 13.12   Let $(X, \leq)$ be a partially ordered set, and let $Y \subseteq X$. An element $b \in X$ is called a **lower bound** of Y if $b \leq y$ for all $y \in Y$. An element $d \in X$ such that $b \leq d$ for all lower bounds b, is called the **infimum** of Y, and is denoted $inf(Y)$. The infimum of Y is sometimes known as the **greatest lower bound** of Y.

The supremum or infimum of a subset may or may not exist. However, if the supremum exists there is at most one supremum. Likewise, if an infimum exists there is at most one infimum. The supremum (or infimum) of a subset Y need not be an element in Y.

Example 13.13
Let $\mathbb{Q}$ be ordered with the standard order. Consider the set X of all positive rational numbers. The infimum of X is 0, note $0 \notin X$. The subset X does not have a supremum.

Example 13.14
Let X be the set $\{a, e, i, o, u\}$. Let $\mathcal{P}(X)$, the power set of X be ordered by subset inclusion. That is for $x, y \in \mathcal{P}(X)$, $x \leq y$ if and only if $x \subseteq y$. Let $S = \{\{a\}, \{a, e\}, \{a, o, u\}\}$. Then, $sup(S) = \{a, e, o, u\}$ and $inf(S) = \{a\}$. So $sup(S) \notin S$ and $inf(S) \in S$.

We say a set is **bounded above** if it has an upper bound and is **bounded below** if it has a lower bound. A set which is both bounded above and below is simply said to be **bounded**.

An important property of the real numbers, $\mathbb{R}$ is that if a non-empty subset of $\mathbb{R}$ has an upper bound then it has a least upper bound. This is not the case for the rational numbers, for instance for $B \subseteq \mathbb{Q}$. where B is the set of all rational numbers less than π. The supremum of B exists in $\mathbb{R}$ but does not exist in $\mathbb{Q}$.

13.4 ENUMERATION AND ISOMORPHISM FOR ORDERED SETS

Sometimes we want to be able to index a set using the natural numbers. That is, we have a partial order $(X, \leq)$ and we want to assign a natural number to each element of X such that the order on X is preserved. This is especially handy when we are attempting to count objects. An injective function $f : X \to \mathbb{N}$ such that if $a \leq b$ then $f(a) \leq f(b)$ is called a **consistent enumeration**.

Proposition 13.15   For any finite partially ordered set $(X, \leq)$ there exists a consistent enumeration.

Proof by Induction:

Let $(X, \leq)$ be a finite partially ordered set with $|X| = n$.

Base Case: $n = 1$. Define $f : X \to \{0\}$ as: for $a \in X$, $f(a) = 0$. Since $|X| = 1$ then by definition, for $a, b \in X$ if $a \leq b$ then $f(a) \leq f(b)$ since $a = b$.

Inductive Hypothesis: Assume that for any partially ordered set $(X, \leq)$ with $|X| = k$ that there exists $f :\to \{0, \ldots, k\}$ such that for $a, b \in X$ if $a \leq b$ then $f(a) \leq f(b)$.

Inductive Step: Let $z \in X$ be a minimal element of X (such an element exists because X is finite). Let $Y = X - \{z\}$. Then $|Y| = k$ and as such there exists $g : Y \to \{0, \ldots, k\}$ which is a consistent enumeration. Consider the function $f : X \to \{0, \ldots, k, k+1\}$ given by $f(z) = 0$ and $f(x) = g(x) + 1$ for $x \neq z$. Since g is a consistent enumeration we have that when $a, b \in Y$, if $a \leq b$ then $g(a) \leq g(b)$ and hence $f(a) \leq f(b)$. Thus, we need only consider a few cases when z is involved. If z and b are comparable for some $b \in Y$ then since z is a minimal element it follows that $z \leq b$. Since $f(z) = 0$ and $f(b) > 0$ for all $b \in Y$ it follows that $f(z) \leq f(b)$. Thus, we have that for all $a, b \in Y \cup \{z\}$ if $a \leq b$ then $f(a) \leq f(b)$ and hence $f : X \to \{0, \ldots, k, k+1\}$ is a consistent enumeration.

Thus, by mathematical induction, the claim follows.　　　　　　　　　　□

One of the main reasons we care about a consistent enumeration, is that it is closely tied to the concept of isomorphism for ordered sets. In mathematics, an isomorphism is a bijection between two structured sets which preserves the abstract structure. For example, as far as cardinality is concerned we simply care about the size of the sets so all that we need to preserve is the size, which bijections already do. Isomorphisms show up in various upper-year mathematics courses and their specifics differ by subject matter but the similarity is that they all preserve some kind of structure. For partially ordered sets we care about the size of the sets but also the relative order. What we don't necessarily care about is the specific names attached to the elements of the partially ordered set.

Definition 13.16　　Let $(X, \leq_S)$ and $(Y, \leq_R)$ be partially ordered sets. A bijective function $f : X \to Y$ which satisfies the condition that: for all $a, b \in X$, $a \leq_S b$ if and only if $f(a) \leq_R f(b)$, is called an **isomorphism**, or more accurately an **order-isomorphism**.

If an order-isomorphism exists between two partially ordered sets those sets are said to be **isomorphic**. What this means is that these partially ordered sets are the same except for how the elements are named, as far as the orderings are concerned. Isomorphic sets are sometimes referred to as **similarly ordered sets** or **similar sets** (when it is clear we are discussing order-isomorphisms rather than simply cardinality).

Example 13.17
Let $X = \{2^k : k \in \mathbb{N}\}$ be the set of powers of two, ordered by the divisibility relation. Let $\mathbb{N}$, the natural numbers, be ordered by the usual ordering on the natural numbers. The function $f : X \to \mathbb{N}$ given by $f(2^k) = k$ is an order-isomorphism.

It is worth noting that inverses and composition preserve the order-isomorphism.

Proposition 13.18   Let X and Y be partially ordered sets. Let $f : X \to Y$ be an order-isomorphism. Then $f^{-1} : Y \to X$ is an order-isomorphism.

Proof:

Let X and Y be partially ordered sets and let $f : X \to Y$ be an order-isomorphism. Then, $\forall a, b \in X$, $a \le b$ if and only if $f(a) \le f(b)$. Thus, $\forall a, b \in X$, $f^{-1}(f(a)) \le f^{-1}(f(b))$ if and only if $f(a) \le f(b)$.

Since f is a bijection, for every $c, d \in Y$ there exists $s, t \in X$ such that $f(s) = c$ and $f(t) = d$. Thus, for all $c, d \in Y$, $f^{-1}(c) \le f^{-1}(d)$ if and only if $c \le d$. Thus, f^{-1} is an order-isomorphism. □

Proposition 13.19   Let X, Y, and Z be partially ordered sets. Let $f : X \to Y$ and $g : Y \to Z$ be order-isomorphisms. Then, $g \circ f$ is an order-isomorphism.

Proof:

Let $f : X \to Y$ and $g : Y \to Z$ be order-isomorphisms. Then, $g \circ f : X \to Z$ is a bijection, since it is the composition of bijections. If $a, b \in X$ and $a \le b$ then since f is an order-isomorphism $f(a) \le f(b)$, and since g is an order-isomorphism $g \circ f(a) \le g \circ f(b)$. □

13.5   ORDER TYPES OF TOTALLY ORDERED SETS

If we have a collection of totally ordered sets and two sets are assigned the same symbol if and only if the sets are order-isomorphic then the symbol is called the **order type** of the sets. In other words, all totally ordered sets which are isomorphic to each other are assigned to the same symbol, which is the order type of those sets.

When θ is the order type of a totally ordered set X, then θ^* is used to denote the order type of the dual order of X.

Some of the common order types are:

ω for the natural numbers, $\mathbb{N}$

π for the integers, $\mathbb{Z}$.

η for the rational numbers, $\mathbb{Q}$.

The idea of an order type will play a roll when we get to the definition of Ordinal Numbers.

13.6   WELL-ORDERED SETS

Now that we have the terminology down, we can define and discuss well-ordered sets.

Definition 13.20   Let X be a partially ordered set. If every non-empty subset of X has a first element (in the induced order) then X is said to be a **well-ordered set**.

That definition may require some explanation. A well-ordered set is a set X where whenever $A \subseteq X$ and $A \neq \emptyset$, then there exists a first element $a \in A$ in the restriction of X to A. In other words, if $A \subseteq X$ and $A \neq \emptyset$ then there exists $a \in A$ such that for all $x \in A$, $a \leq x$.

The first thing to note is that any well-ordered set is a totally ordered set. To see this observe that for a well-ordered set X and any two $x, y \in X$, the set $\{x, y\}$ must contain a first element, so either $x \leq y$ or $y \leq x$ either way x and y are comparable and thus the partial order is a total order.

Proposition 13.21   Let X be a well-ordered set. Every subset of X is well-ordered, and if Y is order-isomorphic to X then Y is well-ordered.

Proof:

This will be a direct proof. Let X be a well-ordered set. Suppose $A \subseteq B \subseteq X$. If $B = \emptyset$ then technically B is well-ordered since B has no non-empty subsets. If $A \neq \emptyset$ then since X is well-ordered and $A \subseteq X$, then A contains a first element. This means that any non-empty subset of B contains a first element and hence B is well-ordered.

If Y is order-isomorphic to X then there exists an order-isomorphism $f : X \to Y$. Let $C \subseteq Y$, with $C \neq \emptyset$, and define $A = f^{-1}(C)$ (define A as the image of the inverse of f). Since X is a well-ordered set, A contains $a \in A$ such that $\forall x \in A$, $a \leq x$. By definition of an order-isomorphism, this implies that $\forall y \in C$, $y = f(b)$ for some $b \in A$ and hence $f(a) \leq y$. Thus, $f(a)$ is a first element of C, which means that every non-empty subset of Y has a first element and hence Y is a well-ordered set.   □

Order-isomorphism preserves the property of being well-ordered. With a bit more terminology we can really understand the types of relationship possible between well-ordered sets.

Definition 13.22   Let X be a partially ordered set. Let $a, b \in X$. Then, b is said to be the **immediate successor** of a if $a \leq b$ and there does not exists $c \in X$ such that $a < c < b$. If b is the immediate successor of a then a is said to be the **immediate predecessor** of b.

Definition 13.23   Let X be well-ordered set. Let $x \in X$ then $s(x)$, the set of all $y \in X$ such that $y < x$, is called the **initial segment** of x.

It turns out that well-ordered sets are very structured, and they have a variety of somewhat surprising properties.

Proposition 13.24   Let X be a well-ordered set and let $A \subseteq X$, with $A \neq \emptyset$. If $f : X \to A$ is an order-isomorphism then $\forall x \in X, x \leq f(x)$.

Proof:

Let $f : X \to A$ be an order-isomorphism. Let $B = \{x : f(x) < x\}$. If B is empty then $\forall x \in X, x \leq f(x)$. The rest of the proof is a proof by contradiction. Suppose $B \neq \emptyset$. Since B is non-empty and X is well-ordered, B contains a first element b s.t. for all $y \in B, b \leq y$. However, $f(b) < b$ by definition of B, thus $f(b) \notin B$. Since f is an order-isomorphism $f(f(b)) < f(b)$ thus by definition $f(b) \in B$. This is a contradiction, thus B is empty and it follows that $\forall x \in X, x \leq f(x)$. $\square$

Proposition 13.24 has some powerful consequences.

Proposition 13.25   Let X and Y be order-isomorphic well-ordered sets. Then, there exists a unique order-isomorphism $f : X \to Y$.

Proof:

Let X and Y be well-ordered sets, and let $f : X \to Y$ and $g : X \to Y$ be order-isomorphisms. Consider $g^{-1} \circ f : X \to X$; Propositions 13.18 and 13.19 tells us that $g^{-1} \circ f$ is also an order-isomorphism. By Proposition 13.24 we have that $\forall x \in X, x \leq g^{-1} \circ f(x)$. Thus, since g preserves order, $g(x) \leq f(x)$ for all $x \in X$. Similarly, $f^{-1} \circ g$ demonstrates that $f(x) \leq g(x)$ for all $x \in X$. Hence, $f(x) \leq g(x)$ and $g(x) \leq f(x)$ for all $x \in X$, by the antisymmetric property for partial orders it follows that $g(x) = f(x)$. Thus, $f = g$. $\square$

Proposition 13.26   Let X be a well-ordered set, and for some $x \in X$, let $s(x) = \{y \in X : y < x\}$ be an initial segment. Then X is not similar to $s(x)$ (i.e., X is not order-isomorphic to $s(x)$).

Proof by Contradiction:

Suppose, by way of contradiction, that $f : X \to s(x)$ is an order-isomorphism. Since $s(x) \subseteq X$, by Proposition 13.24 we have that for all $a \in X, a \leq f(a)$. Thus, $x \leq f(x)$ however, by definition $f(x) \in s(x)$ and so $f(x) < x$ this is a contradiction. $\square$

These results lead to the very concise summary of possible relationships between well-ordered sets.

Proposition 13.27   Let X and Y be well-ordered sets. Either X is similar to Y, X is similar to an initial segment of Y or Y is similar to an initial segment of X.

Sketch of Proof:

Consider the set of all elements of X that have initial segments which are isomorphic to initial segments of Y (and vice versa). Call these sets A and B, respectively, with $A \subseteq X$ and $B \subseteq Y$. Show that A and B are order-isomorphic. Show that $A = X$ or A is an initial segment of X, and similarly $B = Y$ or B is an initial segment of Y, this is the difficult part. These two binary options lead to the four possible cases. The details are left as an exercise.   □

13.7   ORDINAL NUMBERS

Now that we have all the terminology down, ordinal numbers are actually easy to define. Like with order types on totally ordered sets, we assign to every well-ordered set a symbol so that the same symbol is assigned to sets which are order-isomorphic, but different symbols are assigned to well-ordered sets which are not order-isomorphic. For a well-ordered set X the symbol assigned to X is called **the ordinal number** of X. We use the notation $\lambda = ord(X)$ to indicate that λ is the ordinal of X.

We may view ordinals as equivalence classes of all order-isomorphic well-ordered sets. Like with the cardinal numbers, we can discuss **finite** and **transfinite** (or **infinite**) ordinals.

For finite ordinals we will single out one particular representative of a finite ordinal of a given size for convenience. Namely, the one that actually follows the definition of a finite cardinal. Particularly, 0 is used to denote the well-ordered empty set, which is trivially well-ordered. Then for $n \in \mathbb{N}$, $n > 0$, the ordinal of n is the set of all natural numbers which preceed n in the ordering (i.e., $ord(n) = \{0, \ldots, n - 1\}$). This allows us to use the same representation for both finite cardinals and finite ordinals.

Any ordinal that is not a finite ordinal is a **transfinite** or **infinite ordinal**. We reuse the symbols for order types for ordinals. In particular, ω denotes the ordinal of the natural numbers.

Definition 13.28   Define the relation $\leq$ on the *collection* of ordinal numbers as follows: If X and Y are two well-ordered sets and $\lambda = ord(X)$ and $\gamma = ord(Y)$ then let $\lambda \leq \gamma$ if and only if X is order-isomorphic to Y or X is order-isomorphic to an initial segment of Y. This is the *standard ordering of the ordinals*. Note: Proposition 13.27 implies that $\lambda < \gamma$ precisely when X is order-isomorphic to an initial segment of Y.

One consequence of Proposition 13.27 is that Definition 13.28 actually defines a total order on any set of ordinal numbers. In fact, although we will not prove it, this can actually be strengthened to note that the standard ordering of the ordinals in a well-ordering. This means that for any ordinal λ, if we let $s(\lambda)$ denote the set of all ordinal numbers which strictly precede λ in the standard ordering then $s(\lambda)$ is a well-ordered set and hence has an ordinal. The following proposition precisely describes the relationship between λ and $ord(s(\lambda))$.

Proposition 13.29   Let λ be an ordinal number and let $s(\lambda)$ be the set of all ordinal numbers which strictly precede λ in the standard ordering. Then, $\lambda = ord(s(\lambda))$.

It is now clear why, we prefer to use the cardinal definition for finite ordinals, it is actually the natural definition for *any* ordinal. Proposition 13.29 implies that all ordinals have an immediate successor. Namely, if for some well-ordered set X, $\lambda = ord(X)$ then the immediate successor of λ is the ordinal of the set $s(\lambda) \cup X$.

Thus, for instance while $\omega = ord(\mathbb{N})$, and $s(\omega) = \mathbb{N}$, we have that $\omega + 1$ (the immediate successor of ω) is given by the ordinal of the set $s(\omega) \cup \mathbb{N}$.

In fact, we may define addition on ordinals as follows: Let X and Y be disjoint well-ordered sets and let $\lambda = ord(X)$ and $\gamma = ord(Y)$. Then $\lambda + \gamma$ is the ordinal of the set $X \cup Y$ where $X \cup Y$ is ordered by summation order. Note that with ordinal addition $\lambda + \gamma$ need not be equal to $\gamma + \lambda$ (addition is not commutative).

In this fashion we may construct ω, $\omega + 1$, $\omega + 2$, ..., $\omega + \omega$.

Some ordinals, like ω do not have an immediate predecessor. There is a name for these sorts of ordinals. **Limit ordinals** are ordinal numbers which do not have an immediate predecessor. Note that $\omega + 1$ does have an immediate predecessor, namely ω.

13.8   TRANSFINITE INDUCTION

We are already familiar with mathematical induction. We saw in Chapter 6 that mathematical induction rested on the well-ordering principle. The assumption that every non-empty subset of the natural numbers contains a least element. In other words, in the usual ordering the natural numbers is a well-ordered set. It turns out that there is a version of induction for well-ordered sets, which need not be the natural numbers.

At times we may wish to consider, proving statements about well-ordered sets. **Transfinite Induction** is a proof technique that works as follows: Suppose X is a well-ordered set, and we have a predicate $P(x)$, which is true or false for $x \in X$. In order to prove that $P(x)$ is true for all $x \in X$, we do the following: Prove the **base case** when $P(a)$ where a is the first element of X. Next, prove the **successor case** by showing that if the statement $P(x)$ is true for all $b \in X$ which strictly precede $P(b)$ then $P(b)$ must also be true. Note that in this statement of the successor case, we explicitly require that any and all limit ordinal like cases, where an element has no immediate predecessor, must also be dealt with.

The proof technique of Transfinite Induction will be stated in set-theoretic manner which makes it a little less cumbersome to deal with.

Proposition 13.30   Principle of Transfinite Induction
Let X be a well-ordered set, let a be the first element of X, and let $s(x)$ denote the initial segment of x. Suppose Y is a subset of X with the following two properties:

- $a \in Y$

- If $s(x) \subseteq Y$, then $x \in Y$.

Then, $X = Y$.

Proof:

This proof is left as an exercise.

13.9   THE WELL-ORDERING THEOREM AND THE AXIOM OF CHOICE

We have been discussing properties of well-ordered sets but have so far not really been discussing which sorts of sets can, in fact, be well-ordered. It is clear that finite sets can be well-ordered. For countably infinite sets, in particular the natural numbers, we had to adopt the *Well-Ordering Principle* back in Chapter 6. What about sets with a cardinality larger than $\aleph_0$? Cantor's Theorem in Chapter 12 demonstrated that there are an infinite number of such transfinite cardinal numbers.

It would be nice to say the every set can be well-ordered. In fact, many mathematicians believe this to be the case. However, it turns out that this view is somewhat controversial. We need to discuss the *Well-Ordering Theorem*.

Definition 13.31   The Well-Ordering Theorem
Every set can be well-ordered.

Originally the *Well-Ordering Theorem* was demonstrated in 1904 by Zermelo. However, shortly after this proof was published it was discovered that Zermelo had used a property of sets which may actually be deduced by the Well-Ordering Theorem. Thus, the property and the Well-Ordering Theorem are actually logically equivalent. That property, is the *axiom of choice*. We gave a proper definition of the axiom of choice in Definition 12.17, but roughly speaking the axiom of choice states that if we have a collection of pairwise disjoint non-empty sets then a set exists which has a unique element in common with each set in the collection. In other words, given an equivalence relation on sets, it is possible to construct a set made up of single representatives from each equivalence class.

The axiom of choice is problematic for some mathematicians. Some reject the idea that it is possible to make infinitely many selections, equating the use of the axiom to a task which requires an infinite number of steps to finish. Others accept the axiom for countably infinite sets but reject it for sets of larger cardinality. Others accept the axiom of choice and its logical consequences.

Some of the more interesting consequences in set theory are:

- Every set can be well-ordered (The Well-Ordering Theorem).

- Every infinite set contains a subset with cardinality $\aleph_0$.

- Any pair of cardinal numbers are comparable (i.e., the cardinals are totally ordered).

- Any union of a countable collection of countable sets is countable.

- For any set S there exists a bijection $f : S \to S \times S$.

- Any infinite set is similar to a proper subset of itself.

- Let A be a non-empty partially ordered set. If every chain in A has an upper bound in A then, A contains a maximal element. (Zorn's Lemma)

The Well-Ordering Theorem, Zorn's Lemma, and the claim that the cardinals are totally ordered are all logically equivalent to the axiom of choice, some of the other listed results are logical consequences of the axiom. One of the more controvertial consequences is the *Banach–Tarski paradox* a theorem in set-theoretic geometry which roughly states that a ball in three dimensions can be disassembled into a finite number of disjoint subsets and assembled in a different way by rotating and translating the subsets to produce two identical copies of the original ball.

We will return to the axiom of choice in Chapter 14. In the meantime we note that we adopted the axiom of choice for cardinal numbers and continue to accept it.

13.10 ADDITIONAL EXAMPLES

Example 13.32
Suppose X and X' (the dual of X) are both well-ordered. What may be deduced about the cardinality of X? Justify your answer.

Solution:

Suppose X and X' are both well-ordered. We will deduce that $|X|$ must be finite.

The definition of a dual-order means that (a, b) is in the relation for X iff (b, a) is in the relation for X'. Since X' is well-ordered every non-empty subset of X' contains a first element. When this is translated back to X this means that every non-empty set contains a last element. Thus, every non-empty subset of X contains both a first and a last element.

This means that there can be no subsets of X which are order-isomorphic to ω (the natural numbers). Thus, X must have a finite ordinal. Thus, $ord(X)$ is a natural number and hence $|X|$ is finite. $\qquad\square$

Example 13.33
Show that $\omega + 2 \neq 2 + \omega$.

Solution:

Let X be the well-ordered set with $ord(X) = \omega + 2$ given by $X = \mathbb{N} \cup \{a, b\}$ in summation order. Let Y be a well-ordered set with $ord(Y) = 2 + \omega$ given by $Y = \{a', b'\} \cup \mathbb{N}$ in summation order.

The element $b \in X$ is a last element. There are no last elements in Y. Thus, X and Y are not order-isomorphic and hence $\omega + 2 \neq 2 + \omega$. $\qquad\square$

Example 13.34
Prove that there are an infinite number of ordinals of countably infinite cardinality.

Proof:

This will be a direct proof which will assign to each natural number a different ordinal of countably infinite cardinality. First we note that ω is in fact an ordinal of countably infinite cardinality. We will assign ω to the number 0. We will assign to 1 the ordinal $\omega + \omega$, and in general to $n \in \mathbb{N}$ the ordinal $\sum_{k=1}^{n+1} \omega$. Since each $\sum_{k=1}^{n+1} \omega$ can be represented by a finite union of countably infinite sets all of the mentioned ordinals are countably infinite. It remains to show that none of these ordinals are equal to each other.

Notice that in $\mathbb{N}$ there is exactly one element which has no immediate predecessor, namely 0. In $\mathbb{N} \cup \mathbb{N}_1$, the summation order given by gluing two copies of $\mathbb{N}$ together (as long as different symbols are used) there are two elements with no immediate predecessor, namely the two copies of zero. In general for $\sum_{k=1}^{n+1} \omega$ the respresentative well-ordering associated with gluing together

$n + 1$ copies of $\mathbb{N}$ in summation order has $n + 1$ elements with no immediate predecessor, the $n + 1$ copies of 0. Thus, $\sum_{k=1}^{n+1} \omega$ and $\sum_{k=1}^{m+1} \omega$ have different numbers of elements with no immediate predecessor if and only if $n \neq m$, and hence are not order-isomorphic if and only if $n \neq m$. Thus, there are an infinite number of ordinals of countably infinite cardinality. □

13.11 PROBLEMS

Problem 13.35

Let X be a set. Prove that the subset relation is a partial order on the power set of X.

Problem 13.36

Prove that the dual order of a partial order is itself a partial order.

Problem 13.37

Let $(X, \leq)$ and $(Y, \leq)$ be totally ordered sets. Then the **lexicographic order** on $X \times Y$ is the total order defined as: $(x, y) \leq (u, v)$ if and only if, $x < u$ or $x = u$ and $y \leq v$.

Prove that the lexicographic order on $X \times Y$ is a total order.

Hint: You must first prove that it is a partial order.

Problem 13.38

Consider the set $E \cup A \cup (\mathbb{N} - E)$ in Example 13.5. Does it have a first element? Does it have a last element? What about the set $E \cup (\mathbb{N} - E) \cup A$?

Problem 13.39

Let X be a partially ordered set, and let $A \subseteq X$. Prove that if the supremum of A exists then it is unique (i.e., prove that if a and b are both the supremum of A then $a = b$).

Problem 13.40

Let $S = \{x > 1 : x \text{ is an integer }\}$, ordered by the divisibility relation. Which of the following sets are chains in S? For those that aren't explain why they are not.

(i) $A = \{3, 50\}$

(ii) $B = \{3, 6, 12\}$

(iii) $C = \{2, 16, 64\}$

(iv) $D = \{5, 10, 15, 20, 25\}$

(v) $\emptyset$

(vi) $S = \{x > 1 : x$ is an integer $\}$

Problem 13.41

Consider the following relation R defined on the positive integers, $\mathbb{Z}^+$. For $a, b \in \mathbb{Z}^+$, aRb if and only if $b = a^t$ for some integer t.

(i) Prove that R is a partial order on $\mathbb{Z}^+$.

(ii) Prove or disprove that this partial order is a total order.

(iii) Determine the last element in the partial order.

Problem 13.42

Suppose $(X, \leq)$ and $(Y, \leq)$ are partial orders, and that $f : X \to Y$ is an order-isomorphism. Show that for $a \in X$, if $f(a)$ is a first or last element then a is a first or last element, respectively.

Problem 13.43

Suppose $(X, \leq)$ and $(Y, \leq)$ are partial orders, and that $f : X \to Y$ is an order-isomorphism. Show that if $a \in X$ is a minimal or maximal element then $f(a)$ is a minimal or maximal element, respectively.

Problem 13.44

Consider a well-ordered set X whose ordinal is $\omega + 1$. Does it have a first or last element? If so, find them, and demonstrate that they are first or last elements.

Problem 13.45

Let $S = \{1 < x < 6 : x$ is an integer $\}$ ordered by divisibility. Determine three different consistent enumerations $f : S \to \{0, 1, 2, 3, 4, 5\}$.

Problem 13.46

Prove or disprove: $\omega = 1 + \omega$.

Problem 13.47

Recall that in general $a + b \neq b + a$ for ordinals. Prove that if $n, m \in \mathbb{N}$ then, $n + m = m + n$ in ordinal addition.

Problem 13.48

Suppose X is a bounded totally ordered set. Is X well-ordered? Prove your answer.

Problem 13.49

Let A be a partially ordered set and let $B \subseteq A$. If the infimum and supremum of B both exist and are equal what may be deduced about B? Justify your answer.

Problem 13.50

Prove Proposition 13.27.

Hint: Use the provided sketch of the proof as a blueprint.

Problem 13.51

Prove Proposition 13.30, the principle of transfinite induction.

Hint: Look at the proof for the validity of mathematical induction in Chapter 6.

CHAPTER 14

Paradoxes and Axiomatic Set Theory

This chapter discusses some of the paradoxes of naive set theory and the axiomatic set theory approach used to resolve these paradoxes. While this chapter seeks to provide a solid introduction to the subject matter for students first encountering axiomatic set theory it is by no means the most exhaustive or authoritative text. Students interested in a more comprehensive discussion of axiomatic set theory will find *Set Theory and Logic* by Robert R. Stoll an excellent resource.

14.1 PARADOXES OF NAIVE SET THEORY

Naive set theory is the set theory commonly encountered in "everyday" mathematics. It is the notion of set theory we have been discussing for the duration of this entire text. The majority of advanced mathematics uses this intuitive set theory for mathematical theorems and proofs.

However, there are some problems with the theory, which have been called "paradoxes." Strictly speaking, they are actually **antinomies** rather than paradoxes, though very few people actually call them that. In other subject areas, like physics, paradoxes refer to bizarre features which appear superficially to be a contradiction but upon further examination merely highlight a quirk of the theory which is counter intuitive. In intuitive set theory the "paradoxes" *result in actual contradictions*. This is potentially disastrous since as we've already seen with proofs by contradiction *anything* may be proved to be a true claim from a logical contradiction.

The fact that we refer to them as paradoxes should be a hint that all hope for mathematics is not actually lost, and that there is still a silver lining at the end.

14.2 CANTOR'S PARADOX

Let us consider **the set of all sets**, S. Since S contains every set it contains all subsets of S and as such, $\mathcal{P}(S)$, the power set of S, is an element in S. So $\mathcal{P}(S) \subseteq S$, but this implies that $|\mathcal{P}(S)| \leq |S|$. However, according to Cantor's Theorem $|S| < |\mathcal{P}(S)|$. This is a contradiction because it implies that an injective function from $\mathcal{P}(S)$ to S must exist but also cannot exist.

Clearly, this is troubling because the concept of a set of all sets quickly leads to a contradiction. However, it may be that the idea of a set of all sets is simply too big. Perhaps it's too large to use as a universal set. What if we simply decide not to allow such a set?

14.3   RUSSEL'S PARADOX

For Russel's paradox, we consider R the set of all sets which do not contain themselves. Namely, R is defined via set builder notation as $R = \{X : X \notin X\}$. If $R \in R$ then by definition R satisfies the condition $R \notin R$. If on the otherhand, $R \notin R$ then R must be included as a member of R and as such $R \in R$. Thus, we have that $R \in R$ and $R \notin R$.

The problem with Russel's paradox is that we do not even need the universal set to be particularly large. In fact virtually any suitable universal set, that contained R, would yield the same problem. This means that we cannot simply limit our work to a particular domain of discourse and proceed as if everything were fine.

14.4   AN UNSTATED ASSUMPTION

Cantor's paradox demonstrates that some sets are in a sense too big, while Russel's paradox demonstrates that self-referential statements can lead to issues. Underlying both of these is an unstated assumption: *there is a set which corresponds to all objects with a given property.*

In fact, we have been using **the unrestricted principle of abstraction** also known as **the principle of unrestricted comprehension**, the idea that for any given predicate there is a set which satisfies that predicate. We have assumed that $\{x : P(x)\}$ defines a set for any predicate P.

As it turns out, we cannot make that assumption without letting all kinds of paradoxes into set theory. Fortunately, we do have one card left to play.

14.5   AXIOMATIC SET THEORY

Axiomatic set theory was developed to address the shortcomings of naive set theory and provide a solid foundation for mathematics. The paradoxes of naive set theory do not lead to contradictions in axiomatic set theory, however the trade off is that axiomatic set theory is significantly more difficult to work with. So much so that virtually all mathematicians use naive set theory in their proofs with the understanding that as long as certain rules of thumb are followed, it will not be an issue. Nonetheless, it is important to understand some of the ideas of axiomatic set theory, so we will discuss some of the broad strokes. Moreover, it is necessary to begin with naive set theory so that the peculiarities of axiomatic set theory make sense.

14.6   AXIOMATIC THEORIES

In any mathematical subject area, whether it be geometry, set theory, arithmetic or logic, there are certain terms at the very foundation of the subject matter which are understood by intuition. In geometry, these are things like: point, line, plane, congruence, and intersection. In logic, the terms are: proposition, true and false. In naive set theory, we talk about: the empty set, elements, and objects. These foundational terms which are understood by appeals to intuition are called **primitive terms**, also known as **undefined terms**.

In an **axiomatic theory,** or **axiomatic system**, the relations between primitive terms are rigorously defined by axioms, specific rules for how the undefined terms relate to each other. These are axioms are assumed to be true and taken together may be used to derive **theorems**. These theorems are any statement which is *provable* using the axioms of the theory. The axioms together with the theorems which can be logically derived from them are the particular axiomatic theory.

An axiomatic system is **consistent** if it does not contain any contradictions. In other words, axiomatic systems are consistent if and only if it is not possible for a proposition and its negation to both be theorems of the system. An axiomatic system is said to be **complete** if every possible proposition in the system is either a theorem or its negation is a theorem. In other words, a complete axiomatic system is one in which it is possible to prove or disprove every single statement in the system.

We've actually already seen some axiomatic systems. In Chapters 2 and 3, we were introduced to some axiomatic logical systems. Although we did try to describe them in everyday language: propositions, true and false, were all primitive terms and a large amount of intuition is needed to grasp what they mean. The various rules of logic introduced, whether it's the laws of the algebra of propositions or the rules of inference, are all examples of axioms which we used to derive the conclusions of valid arguments. Example 2.20 from Chapter 2 is an example of a proof in an axiomatic system.

Axiomatic systems are used to provide the solid foundation of a subject area. Assuming no human error occurred in the actual proof process, axiomatic systems let us demonstrate that theorems are purely logical consequences of the initial assumptions of the theory. They let us sidestep the ontological issues of mathematics, like in what sense numbers actually exist or whether infinity and reality are actually compatible concepts. They do this by essentially codifying *manipulation rules* for how primitive terms can interact.

Formalism in mathematics takes the view that mathematics is not about a set of propositions which correspond with reality in some abstract way, rather it's more like a game being played according to a set of rules. **Formal axiomatic systems** take this to an extreme. These are axiomatic systems where strings of symbols are manipulated according to specified rules, these rules, along with a set of starting symbols, comprise the axioms of the system. Any statement is a string of symbols, and a formal proof corresponds to a specification of a sequence of rules used to turn a starting string into the final theorem. If that sounds like a computer programming language, you are not wrong. In a sense a programming language is a type of formal axiomatic system.

Formal axiomatic systems can be very difficult for humans to work with, as one needs to manipulate long sequences of symbols according to fixed rules. These days we use computers to do the actual manipulation, to avoid the laborious and tedious process. With these *computer-assisted proofs* we can minimize the chances of human error in the symbolic manipulation, the deductive logical steps of the proof, but we do need to still place our faith in the correctness of the computer program used to manipulate the strings and that it was programmed properly. However, the advantage is that they virtually eliminate all intuitive leaps in a mathematical proof which may not actually be valid.

In order to address the paradoxes of naive set theory, it is necessary to adopt an axiomatic approach. This means that we will need to have *undefined terms*, like set and element. There are actually multiple possible ways to formulate an axiomatic approach to set theory but we will discuss the most commonly used one, which now days in considered to be the foundation of mathematics: Zermelo–Fraenkel Set Theory (with the axiom of choice).

14.7   ZERMELO–FRAENKEL SET THEORY AND THE AXIOM OF CHOICE

The standard axiomatic theory of set theory is Zermelo–Fraenkel Set Theory with the axiom of choice (ZFC). We will discuss its various axioms and comment upon some of the difference between ZFC and naive set theory.

The primitive terms we will begin with are **set** and **membership**. We will be using lowercase letters for sets, because uppercase letters are being used for *formulas*. Membership is denoted $\in$ and is a binary predicate (e.g., $a \in x$). We cannot define membership as a relation because we do not have the notion of what a binary relation is yet. For now we consider $\in$ to symbolically link the symbol which precedes it on the left with the symbol which succeeds it on the right via *membership*.

We do not say what a set is, we are not defining it as a collection of objects. So we must further specify that the symbol that the membership relation precedes be a set. Note, ZFC uses only sets and membership, so that all sets are made up of other (nested) sets or are the null set. However, it is also possible to define **individuals** as symbols which may precede the membership symbol but not succeed the membership symbol. Such individuals are what we thought of as objects which made up sets in naive set theory. For the purposes of this chapter we will proceed without dealing with individuals because ZFC has to be modified substantially to deal with them.

We shall also be including all of the logic from both propositional logic and predicate logic, we note that this means that the equality symbol is included, though we haven't yet specified what it means for sets to be equal, and that we include *formula* such as, $A(x)$ to denote a technical (logical) formula with free variable x.

We did not really get into technical formula yet but these are formula which are made up purely of logical connectives (e.g., $\not$, $\wedge$, $\vee$, or $\rightarrow$), propositions and variables. A primitive formula (not a composite formula) has a **free variable** if the scope of the variable is not defined for the particular occurence of the variable. For example in $(\exists x) P(x, y)$ the x is not free, but the y is. This technical definition is unfortunately necessary to properly understand some of the axioms. With the definition of free variables it's possible to define exactly what we mean by a **statement** or **sentence** in the way we use them in mathematics. A **statement** is a formula in which there are no free variables. E.g., $(\forall x)x > 1$ is a statement but $(\exists y)x = y$ is not (since x is a free variable).

Axiom 14.1   Axiom of Existence (ZF0)
There exists a set.

Axiom 14.2   Axiom of Extension (ZF1)
If x and y are sets, and $\forall a$ we have $a \in x$ iff $a \in y$ then $x = y$.

Axiom 14.3   Axiom Schema of Specification (ZF2)
Let $P(z)$ be a formula which does not contain any occurrence of x or y as free variables.

For any set x there exists a set y such that $\forall a$, $a \in x$ iff $a \in y$ and $P(a)$.

The Axiom of Existence (ZF0) is not usually listed as its own separate axiom, because a later axiom will imply the existence of at least one set, but it is simply much more convenient to do so now so that we can show that the empty set also exists. The Axiom of Extension (ZF1) defines the equality of sets we are familiar with, in a formal manner. The axiom schema of specification (ZF2) is actually an infinite number of axioms, one for each choice of formula $P(z)$. ZF2 gives us a way of constructing subsets given an initial set. It replaces the *unrestricted principle of abstraction*. We cannot specify any set by simply defining it as all elements which satisfy a property but we can specify a subset of a set which has a property.

ZF0 and ZF2 together imply that the **empty set** exists. To see this, note that using $x \neq x$ as the formula $P(x)$ in ZF2 which yields the existence of the null set. We adopt the familiar notation $\emptyset$ to refer to the unique null set (uniqueness is guaranteed by the axiom of extension). Also note that, since intersection may be defined purely using the notions of logical conjunction, set membership and ZF2 we adopt the familar $\cap$ notation for the intersection of sets. This version of intersection behaves exactly the same as in naive set theory.

Axiom 14.4   Axiom of Regularity (ZF3)
For every non-empty set x, there exists $a \in x$ such that $a \cap x = \emptyset$.

The Axiom of Regularity (ZF3) states that no set may be an element of itself. ZF3 (along with ZF2) specifically address Russel's Paradox. No set may contain itself so the idea of the set of all sets which do not contain themselves becomes equivalent to the idea of a set of all sets.

Also of note, ZF2 yields the following for any set a, the set $b = \{x \in a : x \notin x\}$ exists, and $y \in b$ iff $y \in a$ and $y \notin y$. What is interesting is that $b \notin a$. The proof is by contradiction and is left as an exercise. What this means is that since a was an unspecified set, we can show that there is always a set b which is not contained in a. In other words, there is no set which contains every set.

It should be noted that these axioms can all (and indeed need to) be stated as formal sentences. For instance the axiom of existence is simply: $(\exists x)$ while the axiom of regularity is: $(\forall x)(x \neq \emptyset \rightarrow (\exists a \in x, a \cap x = \emptyset))$. It is simply easier to understand when written in plain language.

Since ZF2 is used instead of the unrestricted principle of abstraction, we need alternate ways of constructing sets. We can use ZF2 to construct subsets from sets but so far we are only guaranteed that the empty set exists (we know a set exists and that if a set exists then so does the empty set). We move on to axioms used to *carefully* construct larger sets from given sets.

Axiom 14.5   Axiom of Pairing (ZF4)
If x and y are sets then there exists a set z such that $x \in z$ and $y \in z$.

ZF2 along with ZF4 implies that given sets x and y, two element sets of the form $\{x, y\}$ exist, and are called **pairs**, and single element sets of the form $\{x\}$, called **unit sets**, exist. To see this note that "$a = x$ or $a = y$" can be used as the predicate formula in ZF2 and ZF4 thus guarantees that the existence of $\{a \in z : a = x \text{ or } a = y\}$ may be deduced. Since we may construct sets of the form $\{a\}$, $\{\{a\}\}$, $\{\{\{a\}\}\}$ and the like, at this point we could actually construct sets corresponding to the natural numbers. For example, starting with the empty set, let the empty set correspond to zero and have the successor of a natural number be the unit set which contains the preceding natural number.

We can now see one of the reasons why it is that naive set theory is employed in everyday mathematics ... it's extremely annoying to have to constantly specify the correspondence of ZFC sets to the objects that we actually want to discuss. In ZFC sets are essentially nested brackets, it is all structure, so we need to specify what each structure corresponds to.

While it was easy enough to specify intersection (with ZF2 and logic), we actually need a separate axiom for the union of sets. This statement may seem bizarre but the following is the standard way of actually specifying union.

Axiom 14.6   Axiom of Union (ZF5)
For every set z there exists a set x s.t. if $a \in y$ for some $y \in z$ then $a \in x$.

This axiom may seem confusing but we need to keep in mind that sets are made up of sets. Using the previous axioms we can reduce ZF5 to saying that for any sets x and y the union of the set $\{x, y\}$ exists, denoted $x \cup y$, and it contains all sets which were contained in x or y.

With union and intersection, it's now possible to define the *relative complement* and *symmetric difference* of sets in the usual manner: $x - y = \{a \in x : a \notin y\}$ and $x + y = (x - y) \cup (y - x)$, respectively.

The last piece of major definitions from naive set theory that we need are *ordered pairs*. Binary relations, functions, notions of cardinality etc. all rely on the ability to construct ordered pairs. For sets x and y we may construct the unit set $\{x\}$ and the pair $\{x, y\}$ and from these the set which we denote $< x, y >= \{\{x\}, \{x, y\}\}$, and call the **ordered pair** $< x, y >$.

Axiom 14.7   Axiom of Power Set (ZF6)
For every set x there exists a set z s.t. for all a, if $a \subseteq x$ then $a \in z$.

The usual definition of subset applies. The Axiom of Power Set specifies that the power set of any set is also a set. Using a tedious and long formula we can use the Axiom of Power Set to show that the *Cartesian product* of sets a and b is a set. From there we can define a *binary relation* as a set whose members are all ordered pairs (or equivalently as members of the power set of a Cartesian product). For here it's possible to construct definitions for functions and relations corresponding to those found in Chapters 7 and 10.

While we've been able to reconstruct a large portion of the material for naive set theory using ZF0–ZF6, including virtually anything to do with finite sets of arbitrarily large size we note that power sets and unions of finite sets only ever yield finite sets, so we cannot construct an infinite set.

Axiom 14.8   Axiom of Infinity (ZF7)
There exists a set z s.t. $\emptyset \in z$ and if $x \in z$ then $x \cup \{x\} \in z$.

The **successor** of a set x, denoted x^+, is defined as $x^+ = x \cup \{x\}$. The Axiom of Infinity builds in the idea of a successor set, and assigning $\emptyset$ to the natural number 0 in order to specify the existence of the set of natural numbers $\mathbb{N}$. The axiom of infinity is necessary to construct countably infinite sets, which are used to construct the real numbers, as well as other uncountably infinite sets. Subject areas like calculus deal with real numbers and uncountably infinite sets and so definitely require the Axiom of Infinity (or an equivalent axiom), but it is possible to deal exclusively with finite numbers in some areas of math, such as parts of combinatorics. There are versions of axiomatic set theory which do not use the Axiom of Infinity, but it is adopted as standard for the vast majority of mathematics.

The axioms mentioned so far (including ZF7) allow us to actually reconstruct the majority of set theory outside of the theories of cardinals and ordinals. The next axiom is a somewhat controversial one. The Axiom of Choice may seem reasonable, but as mentioned in Chapter 13, it has some far reaching consequences. For instance, the proposition that a surjection from a set A to B implies the existence of an injection from B to A appears to actually require the Axiom of Choice.

Axiom 14.9   Axiom of Choice (ZF8)
For any non-empty set x there exists a function $f : (\mathcal{P}(x) - \emptyset) \to x$ such that for all $y \subseteq x$ with $y \neq \emptyset$, $f(y) \in y$.

The Axiom of Choice states that for any set x there exists a *choice function* which assigns to each non-empty subset of x an element in that subset. While it's easy to see why this would be the case for finite sets, or even countably infinite sets, it isn't as clear for uncountably infinite sets. This is because, as mentioned before, the Axiom of Choice (or equivalently the Well-Ordering Theorem) is significantly more powerful than the Well-Ordering Principle.

We previously defined an infinite set as a set which does not have a finite cardinal number. Proposition 12.7 states that no finite set can be similar to a proper subset of itself. It turns out that with the Axiom of Choice we can define infinite sets in the more elegant form: *a set is infinite if and only if it is similar to a proper subset of itself.*

The next axiom schema is included in ZFC in order to ensure that sets can "be defined" as the image of a set under a well-defined mapping.

Axiom 14.10   Axiom Schema of Replacement (ZF9)
If $S(a, b)$ is a formula s.t. $\forall a$ in a set x, $S(a, b) \wedge S(a, c) \to (b = c)$ then $\exists$ a set y s.t. $b \in y$ iff $\exists a \in x$ s.t. $S(a, b)$.

The final piece of ZFC necessary to reconstruct (a slightly modified version of) Cantor's transfinite cardinals is ZF9. In less formal terms ZF9 states, if given a set x there is some formula

$S(a, b)$ which relates each $a \in x$ to a unique set b, then there is a set y which contains exactly those elements b which the elements of x are uniquely related to. From a practical point of view this means we can define sets in two new ways: as the image of a function (defined without a range set) or by switching members of one set with members of another set, provided a formula exists which can describe the switch. So we could, for instance use the existence of the infinite set from ZF7 and a well-defined formula to demonstrate the existence of sets such as: $\{x, \mathcal{P}(x), \mathcal{P}(\mathcal{P}(x)), \ldots, \mathcal{P}^n(x), \ldots\}$.

14.8   REMARKS ON AXIOMATIC SET THEORY VS. NAIVE SET THEORY

As mentioned earlier, most mathematicians use naive set theory rather than axiomatic set theory for everyday use. It is significantly easier and more intuitive to mentally reason about and discuss sets if they contain identifiable objects as elements. The known paradoxes of naive set theory all concern *self-reference* or *the unrestricted principle of abstraction*. As long as the mathematician is not actually dealing with either issue or attempting to determine whether or not certain statements can be proved in a particular axiomatic system, there isn't really any particular need to deal with ZFC explicitly. ZFC was carefully built to reconstruct as much of naive set theory as is salvageable and does quite a good job at that.

ZFC is the standard axiomatic version of set theory, as well as the standard foundation for general mathematics. However, there are a variety of axiomatic systems equivalent to ZFC which use slightly different axioms. There are also just plain different axiomatic systems which do not try to reconstruct all of ZFC or introduce more controversial axioms which may at times be bizarre. However, there are two very important results which apply to virtually all axiomatic systems that sketch the boundaries of what axiomatic systems are actually capable of.

14.9   GÖDEL'S INCOMPLETENESS THEOREMS

There are two famous mathematical results, published by Kurt Gödel in 1931, called **Gödel's Incompleteness Theorems**, which demonstrate the limitations of almost every axiomatic system. These results only apply to formal systems which are complex enough to express the arithmetic of the natural numbers (not, for instance, systems like predicate logic from Chapter 3).

Recall that we don't really have any way to say whether a statement in an axiomatic theory is "true" or "false" other than showing that it is or is not a logical consequence of a set of axioms. So, the only real notion of "truth" in an axiomatic system is whether a proof of the statement exists. From a formalist point of view this means that the only notion of "truth" in an axiomatic system is whether it is possible for an algorithm to prove the statement.

Gödel's first incompleteness theorem basically states that if a system is consistent and recursively enumerable (an algorithm could theoretically list every theorem of the theory without listing a non-theorem) then there are truths about the arithmetic of the natural numbers that the theory cannot prove. So, there are true statements about the natural numbers that no consistent theory can prove or disprove. In other words, there is no axiomatic system that is the foundation for *all* mathematics.

While the initial reaction may be that, all *useful* theorems are provable in ZFC, there are actually some legitimately interesting questions which cannot be answered using ZFC (i.e., they are independent of ZFC). One of those is the **continuum hypothesis**: There is no set with a cardinality strictly between $\aleph_0$ and $\aleph$. In other words, there's no way to determine whether there is an infinity between the natural numbers and the real numbers in ZFC.

Devastating as this result is, the second incompleteness theorem may seem even more disheartening. Gödel's second incompleteness theorem showed that any consistent formal system cannot prove it's own consistency. In other words, no matter which axiomatic system you choose, you cannot prove it is consistent, using the axioms and logic of the system, unless it is in fact inconsistent (in which case you can prove anything).

Before we leap to the conclusion that axiomatic systems are useless, keep in mind two things. One, these results do not apply to systems of predicate logic (like the type we encountered in Chapter 3), and in fact Gödel proved predicate logic is complete and consistent. Thus, everyday logic has a firm foundation. Two, the second theorem only states that an axiomatic systems consistency is not provable, in that same system. Oddly enough, this isn't actually bad news. A proof for consistency isn't all that desirable. Note that an inconsistent axiomatic system can prove anything so it could prove it's own consistency. This means that, even if the second theorem weren't true, the existence of a consistency proof in a system couldn't actually guarantee that the system is consistent.

With that discussion out of the way we can see that no matter the choices of axioms, any axiomatic system we use for the foundations of mathematics won't allow us to prove all true statements and also cannot guarantee consistency. This means we will always need to evaluate the appropriateness of the axioms as the primary method of guaranteeing "truth" (in the sense of correspondence to reality). This is good news for mathematicians, there will always be more theorems to prove in a different axiomatic system.

14.10 ADDITIONAL EXAMPLES

Example 14.11
Prove that the Well-Ordering Theorem implies the Axiom of Choice.

Proof:

Suppose the Well-Ordering Theorem holds. Let X be any set. Then X can be totally ordered according to some ordering. We pick one such ordering for X. Define a function f as follows: if Y is a non-empty subset of X, $f(Y)$ is assigned the first element of Y according to the total ordering for X. The function f satisfies the definition of the choice function from the Axiom of Choice.   □

Example 14.12
Curry's paradox is a logical paradox where an arbitrary claim can be proved from a self-referential sentence. For example, "If this sentence is true then goats can fly." Self-reference can still be a problem even in formal logic.

Show that the premise $X \equiv (X \to Y)$ implies Y.

Solution:

We will use the tautology $\neg X \vee X$ and the ability to substitute X for $(X \to Y)$ in a rules of inference style proof.

1. $X \equiv (X \to Y)$ (Premise)
2. $\neg X \vee X$ (Premise)

3. $\neg X \vee (X \to Y)$ (by substitution of 1. into 2.)
4. $\neg X \vee (\neg X \vee Y)$ (by definition of implication on 3.)
5. $\neg X \vee Y$ (by idempotent law on 4.)
6. $X \to Y$ (by definition of implication on 5.)
7. X (by substitution of 1. into 6.)
8. Y (by Modus Ponens on 6. and 7.)

14.11 PROBLEMS

Problem 14.13
The liars paradox is the paradox arising from statements of the form "This sentence is false." or

"I am lying." Clearly, these are declarative statements, yet they seem to lead to contradictions. Can you reconcile the liars paradox with propositional logic?

Problem 14.14

Test you heroic decision-making with the following: An insane supervillain, Dr. Paradoxical, holds a studio audience hostage. Dr. Paradoxical states that he will let the hostages go unharmed if and only if you can correctly predict what he will do next. In order to buy yourself time to act, what should you predict?

Problem 14.15

Consider the relation Q on the real numbers in which $(a, b) \in Q$ if $a - b$ is a rational number. First, prove that Q is an equivalence relation. Second, compute the cardinality of the equivalence class of zero.

Problem 14.16

Adopt the setup in Problem 14.15 and consider the equivalence classes under the relation Q. The Axiom of Choice says that there is a set S that contains one number from each of these equivalence classes. If $D = \{u - v : u, v \in S\}$ then prove there is only one rational number in D. Compute the cardinality of S.

Problem 14.17

Using ZFC: For any set a, the set $b = \{x \in a : x \notin x\}$ exists. Prove that $b \notin a$.

Hint: Try a proof by contradiction.

Problem 14.18

Consider ZF without the Axiom of Infinity. Describe the sets in the simplest such axiomatic system.

Problem 14.19

Show that the set of all cardinal numbers leads to a contradiction in naive set theory.

Problem 14.20   Show that the set of all sets similar to the set of natural numbers leads to a contradiction in naive set theory.

Bibliography

Gonick, Larry. *Brains and Bronze (Cartoon History of the Universe vol. 5)*, Rip Off Press, 1980. 169

Mach, Ernst. *Popular Scientific Lectures, translation by T. J. McCormack*, The Open Court Publishing Co., Chicago, 1898. 119

Pasteur, Louis. *Lecture*, University of Lille, 1854. 48

Stoll, Robert R., *Set Theory and Logic*, Dover, 1979.

Weber, H. Leopold Kronecker, *Mathematische Annalen*, 43:1–25, Springer, 1893. DOI: 10.1007/bf01446613. 105

Authors' Biographies

DANIEL ASHLOCK

Daniel Ashlock was awarded a Ph.D. in Mathematics from the California Institute of Technology. He has taught more than 50 different classes from abstract algebra to bioinformatics but enjoys set theory because it is the first time most students meet abstract mathematics. Dr. Ashlock is a Professor of Mathematics at the University of Guelph where he uses mathematics to help biologists with their research as well as continuing his own work in how to represent information to make it easier to understand and easier to work with on a computer. Dr. Ashlock is a Senior Member of the IEEE and has chaired technical committees in both bioinformatics and games, demonstrating the broad usefulness of an education in mathematics.

COLIN LEE

Colin Lee received his Ph.D. in Applied Mathematics from the University of Guelph. While at the University of Guelph he taught set theory both initially as a teaching assistant for Dr. Daniel Ashlock, as well as subsequently as a lecturer in his own right. Since graduating he has been sporadically involved in mathematical research, education, and tutoring while pursuing a variety of endeavours including consulting, programming, commercial art, and authoring mystery and horror fiction.

Index